Sustainable Energy in China
The Closing Window of Opportunity

Noureddine Berrah
Fei Feng
Roland Priddle
Leiping Wang

THE WORLD BANK

©2007 The International Bank for Reconstruction and Development / The World Bank
1818 H Street, NW
Washington, DC 20433
Telephone: 202-473-1000
Internet: www.worldbank.org
E-mail: feedback@worldbank.org

All rights reserved

1 2 3 4 10 09 08 07

This volume is a product of the staff of the International Bank for Reconstruction and Development / The World Bank. The findings, interpretations, and conclusions expressed in this volume do not necessarily reflect the views of the Executive Directors of The World Bank or the governments they represent.
 The World Bank does not guarantee the accuracy of the data included in this work. The boundaries, colors, denominations, and other information shown on any map in this work do not imply any judgement on the part of The World Bank concerning the legal status of any territory or the endorsement or acceptance of such boundaries.

Rights and Permissions
The material in this publication is copyrighted. Copying and/or transmitting portions or all of this work without permission may be a violation of applicable law. The International Bank for Reconstruction and Development / The World Bank encourages dissemination of its work and will normally grant permission to reproduce portions of the work promptly. For permission to photocopy or reprint any part of this work, please send a request with complete information to the Copyright Clearance Center Inc., 222 Rosewood Drive, Danvers, MA 01923, USA; telephone: 978-750-8400; fax: 978-750-4470; Internet: www.copyright.com.
 All other queries on rights and licenses, including subsidiary rights, should be addressed to the Office of the Publisher, The World Bank, 1818 H Street, NW, Washington, DC 20433, USA; fax: 202-522-2422; e-mail: pubrights@worldbank.org.

DOI: 10.1596/978-0-8213-6753-8

Library of Congress Cataloging-in-Publication Data

Sustainable energy in China: the closing window of opportunity / Noureddine Berrah ... [et al.].
　p. cm.
　Includes bibliographical references and index.
　ISBN-13: 978-0-8213-6753-7
　ISBN-10: 0-8213-6753-6
　ISBN-13: 978-0-8213-6754-4 (electronic)
　ISBN-10: 0-8213-6754-4 (electronic)
 1. Energy industries—Environmental aspects—China. 2. Energy policy—Environmental aspects—China. 3. Energy development—Environmental aspects—China. I. Berrah, Noureddine. II. World Bank.
　HD9502.C62S868 2007
　333.790951--dc22
　　　　　　　　　　　　　2007004240

Contents

Foreword   xiii
  by Qingtai Chen
Foreword   xv
  by David Dollar
Foreword   xvii
  by Christian Delvoie
Preface   xix
About the Authors   xxiii
Acknowledgments   xxv
Acronyms and Abbreviations   xxix
Executive Summary   xxxi
  The Problems China Faces   xxxii
  Sustainability: A Challenging but Feasible Goal   xxxvi
  The Path to Sustainability   xxxix
  Sequencing the Steps to Sustainability   xlvii

Chapter 1	**Introduction**	1
	An Impressive Foundation of Past Achievements	1
	Growing Concerns about China's Energy Future	2
	The Four Pillars of Energy Sustainability	5
	The Closing Window of Opportunity	6
	Structure of the Report	7

Chapter 2	China's Energy Future:	
	The Challenge of Recent Trends	11
	Sustained Consumption Growth, 1980–2000	13
	High Growth Forecast for 2000 to 2020	19
	Recent Signs of an Unsustainable	
	Energy Growth Path	22
	The Urgency for Policy Action	30
Chapter 3	Reining in Future Energy Consumption	35
	Achievements in Energy Efficiency	37
	Concerns about Energy Consumption Trends	
	of the 10th Five-Year Plan	41
	Bold New Directions of the 11th Five-Year Plan	41
	The Closing Window of Opportunity	
	for Reducing Long-Term Energy Intensity	43
	The Missing Link: An Improved Policy	
	and Institutional Framework	63
Chapter 4	Greening the Energy Sector	67
	Pollution Levels Still a Major Concern	69
	Environmental Impacts of the Energy Scenarios	72
	The Energy Path to Greener Development	74
Chapter 5	Securing Energy Supply	87
	China's Growing Sense of Insecurity	89
	Options for Securing Oil and Gas Supply	96
	Options for Securing Electricity Supply	115
	Choosing the Right Mix and Amount	
	of Energy Supply Security Measures	119
Chapter 6	Getting the Fundamentals Right	127
	Furthering the Reform Agenda	
	and Developing a Sound Pricing Framework	129
	Energy Pricing in Competitive Markets	131
	Setting Sound Regulated Tariffs	133
	Taxing Energy Commodities	136
	Measures to Mitigate the Environmental	
	Effects of Energy Use	138
	Energy Commodity Price Policy:	
	Status and Necessary Changes	141

Contents vii

Chapter 7	Shaping the Future toward Sustainability	149
	At the Threshold of Change toward Sustainability in Energy Structure and Policy	151
	Characteristics of a Comprehensive Policy for Energy Sustainability	154
	Four Guiding Pillars for Sustainability Policy	156
	Building Blocks to Put the Energy Sector on a Sustainable Path	157
	The Sequence of Steps to Sustainability	170
Appendix A	Gross Domestic Product and Energy Consumption in China, 1980–2005	181
Appendix B	Biomass Energy Use in China	187
	Summary	187
	Biomass Resources	187
	Current Status of Development	188
	Prospects for Biomass in China's Energy Balance	188
	Adverse Effects on the Natural Environment and on Human Health from Biomass Use	189
	Policies Designed to Encourage the Use of Modern Biomass Technologies	190
Appendix C	The Chinese System for Energy Statistics: History, Current Situation, and Ways to Improve the System	192
	The Groundwork Established in the 1980s	192
	Retrenchment in the 1990s	193
	The Current Situation and Its Weaknesses	193
	Conclusions and Recommendations	194
Appendix D	Energy Costs as a Proportion of Gross Domestic Product: Estimates for China, Japan, and the United States	198
Appendix E	Feedback from the Dissemination Workshop	201
	Feedback 1: China's Energy Economy Presents Many Challenges	201
	Feedback 2: The Energy Statistical System Needs Revision	202

	Feedback 3: The Target 20 Percent Energy-Intensity Reduction during the 11th Five-Year Plan Will Be Difficult to Achieve	203
	Feedback 4: Economic Well-Being, Energy Consumption, and Environmental Concerns Should Be Addressed	204
	Feedback 5: Energy Technology Leapfrogging Requires Clarification	205
	Feedback 6: International Cooperation and Technology Transfer Are Needed	206
Appendix F	**Life-Cycle Costs of Electricity Generation Alternatives with Environmental Costs Factored In**	**208**
Appendix G	**International Experience of Insecurity of Energy Supply**	**212**
	Concerns about Oil Imports	212
	Gas Supply Concerns	219
	Concerns about Electricity Supply Failures	224
	Coal Supply, Competitiveness, and Environmental Concerns	226
Appendix H	**Strategic Oil Reserves for China**	**228**
	International Practice	228
	Commercial Stocks	229
	Rationale for Stockpiling	229
	Rationale for a Particular Level of Stocks	230
	Decisions to Draw Down Stocks	230
	Costs	230
	A Strategic Oil Reserve for China?	231
	Quality of Estimates	232
Appendix I	**Predominant Approaches for Setting Regulated Tariffs for Gas and Electricity Transmission and Distribution**	**233**
	Period between Full Tariff Cases	234
	Three Simple Rules Necessary for Determining Tariffs	234
	Revenue Requirements	235

	Rules for Determining Tariffs	238
	Tariff Adjustments between Major Tariff Proceedings	239
	Concluding Comments	239
Appendix J	**Lessons from International Experience: Relevant Examples of Losses Derived from Unsound Energy Pricing**	**241**
	Crude Oil and Products	243
	Natural Gas	243
	Electricity	244
	Effects on General Economic Performance of Nonmarket Energy Pricing	244
	Conclusions	246
Appendix K	**Gas Price Formation and Gas Subsector Reform**	**248**
	Market Pricing Works	248
	Market Prices Can Work for Gas	248
	The Current Gas Pricing System Is Not Tenable	249
	The Target: Wholesale Competition	249
	Transition Could Be Phased	249
	Anticipated Outcomes	254
Appendix L	**Pricing System to Support Adequate Implementation of State Council Document No. 5 on Power Subsector Reform**	**256**
	Bid-Price Pool	258
	Environmental Costs	258
	Transmission Pricing	258
	Final Consumer Pricing	259
Index		261

Boxes

1.1	The Evolving Definition of Energy Security	3
2.1	China's Coal Production Prospects through 2020	28
3.1	The Potential Benefits of Technical Retrofitting: The Case of the Jinan Iron and Steel Group Corporation	54
3.2	Reconciling Automotive Industrial Policy and Energy Sustainability	58

3.3	Energy Efficiency in Buildings	60
4.1	Key Elements for Environmental Improvement in the Green Growth Scenario	73
4.2	China's Clean Development Fund: Leveraging Carbon Finance for Technology Transfer	82
5.1	Bank Involvement in China's Coal Subsector	94
5.2	Examples of Domestic Energy Development Incentives in Other Industrial Countries	98
5.3	Repatriating Equity Oil: Is There a Better Way to Provide Security?	102
5.4	The IEA's Approach to Short-Term Oil Emergencies	111
5.5	The Shell Group's Latest Energy Security Scenarios	113
5.6	Strengthening China's Energy Security: China's Special Concern	120
7.1	Effective Energy Institutions Involve Significant Staffing and Budgets	152
7.2	The U.S. Energy Policy Act of 2005	160
7.3	Key Elements of a Program for a 20 Percent Improvement in Energy Efficiency	162
7.4	Creation of a Strong Energy Ministry Reflects International Practice	164
7.5	China and the International Energy Agency	167
G.1	The 2003–06 Oil Price Spike—How Significant Is It in the Security Debate?	216
G.2	Causes of the First 21st Century Oil Shock	217
G.3	Regional and Liquefied Natural Gas Supply Issues	220
G.4	Canadian Views on the Security of Gas Supply	223
J.1	Impacts of Selected Failed Policies	242
J.2	The Present Gas Supply-and-Demand Situation in North America	245

Figures

2.1	Energy Intensity and Energy/GDP Elasticity, 1980–2005	16
2.2	Net Oil Imports, 1990–2005	27
2.3	GDP per Kilogram of Oil Equivalent of Energy Use	31
3.1	China's Per Capita Energy Consumption and GDP Compared with Selected Countries and the World Average	44
3.2	China's Projected Growth Path of Energy Demand, 1980–2020, Compared with That of Other Countries	45

3.3	China's Energy Intensity and Major Development Periods, 1954–2005	46
3.4	Tunneling a Less Intensive Energy Path to Higher Per Capita Income	47
3.5	Energy-Intensive Industries to Double Output by 2020	53
3.6	Projected Truck Fleet by 2020	55
3.7	Increase in Vehicle Population	56
4.1	Expected Growth in Global Carbon Dioxide Emissions through 2020	72
4.2	Funding the Technological Leapfrogging	84
5.1	Energy Insecurity: Generic Causes, Effects, and China's Special Concerns	90
7.1	Designing and Implementing a Coordinated Energy Policy	158
F.1	Levelized Cost Comparison of IGCC and Subcritical 600-Megawatt with FGD Units: Capital Cost Ratio	210
F.2	Levelized Cost Comparison of IGCC and Subcritical 600-Megawatt with FGD Units: Change of Capital Cost of IGCC	211
I.1	Three Simple Rules for Determination of Tariffs	235
K.1	The Gas Supply Chain: Structure, Contracting, and Pricing Prior to Introduction of Wholesale Competition	250
K.2	The Gas Supply Chain: Structure, Contracting, and Pricing after the Introduction of Wholesale Competition	255

Tables

2.1	Primary Energy Production and Consumption, 1980–2000	14
2.2	Final Energy Consumption by Economic Sector and Fuel, 1980–2000	18
2.3	Key Policy Elements Affecting the Projections of the DRC and ERI's Scenarios for 2020	20
2.4	Projections of China's Primary Energy Consumption, 2000–20	21
2.5	Final Energy Consumption by Sector and Fuel, 2000–20	23
2.6	China's Primary Energy Production and Consumption, 2000–05	24
2.7	China's Oil Consumption and Trade, 1990–2005	26
3.1	Change in China's Energy Intensity by End-Use Sector, 1980–2000	38

3.2	Improvements in the Efficiency of Key Energy-Using Equipment and Comparison with International Standards, 1980–2000	39
3.3	Sectoral Composition of China's GDP, 1980–2000	40
3.4	Energy Demand/GDP Elasticities of Major Industrial Countries, 1961–2002	47
3.5	Household Appliances per 100 Households, 1995–2002	62
4.1	Projections of Key Air Pollutants in the Energy Scenarios for 2000–20 and Government Caps	74
5.1	Oil and Gas Reserves in the Middle East and Russian Federation	99
5.2	Assessment of Relative Difficulty, Cost, and Degree of Control of Key Measures for Short-Term Oil Supply Security	104
6.1	Fuel Taxation in Selected Countries	137
6.2	A Classification of Market-Based and Regulatory Instruments	140
7.1	Shaping the Future: The Issues, Guiding Pillars, Building Blocks, and Sequencing Steps to Sustainability	176
B.1	Development of Biomass Energy in China	189
C.1	Final Energy Consumption by Sector, 1980–2000	195
C.2	Final Energy Consumption by Fuel Type, 1980–2000	196
C.3	Conversion Factors for Tables C.1 and C.2	196
D.1	Total Primary Energy Consumption and Fuel Shares in China, 2005	199
D.2	Total Primary Energy Consumption and Fuel Shares in Japan, 2005	199
D.3	Total Primary Energy Consumption and Fuel Shares in the United States, 2005	200
F.1	Main Technical Indices of IGCC and Subcritical 600-Megawatt with FGD Units	209
G.1	Major Global and Regional Oil Supply Crises since 1950	213
G.2	Some Major Electricity Failures of the Past 40 Years	225
K.1	A Phased Approach to the Introduction of Wholesale Competition	251
L.1	New Tariff Requirements	259

Foreword

Since the reform programs in the 1980s, China's energy industry has witnessed rapid development and increased productivity. China has produced as much as 94 percent of its own energy supply; its production of major energy products has provided powerful support to the attainment of the country's ongoing sustainable social and economic development program. In this same period, China achieved significant progress in improving its energy efficiency. From 1980 to 2000, China's gross domestic product (GDP) increased fourfold, while energy consumption only doubled.

More recently, however, the expansion of the heavy and chemical industries, the acceleration of urbanization and modernization, and the expeditious shift of the international manufacturing industry to China have resulted in substantial increases in total energy consumption, from 1.39 billion tons of coal equivalent (tce) in 2000 to 2.25 billion tce in 2005. In addition, the elasticity of energy consumption to GDP has increased from 0.5 during 1980–99 to more than 1 during 2000–05. Both the increase in energy consumption and the decrease in energy efficiency have exerted adverse environmental effects and China became one of the major SO_2 emitters in the world.

China's development faces a strategic crossroads. On the one hand, by 2020 China will achieve the GDP target of fourfold growth over that of 2000 established by its long-term social and economic development plan. Such growth will generate substantial new investment. The energy consumption and pollutant emission levels of the asset increments will have negative long-term effects. If necessary new energy policies are not effectively integrated into a supportive legislative framework, and if measures to reduce the energy consumption intensity and enhance environmental protection are not implemented, any development will be accompanied by high consumption, severe pollution, and low efficiency. If this happens, China would confront high prices and unsustainable development.

On the other hand, China has a historic opportunity to achieve sustainable development. The rapid economic development and increased market demand place China in an advantageous position to borrow international advanced practices in the new energy capacities, promote technical innovations, and achieve an updated energy structure. We must take this opportunity to design and implement national long-term energy development strategies and policies that are based on a thorough study of China's energy and environmental issues.

China's energy issue has generated international attention, resulting in numerous productive discussions and deepening collaborations. For many years, the World Bank has supported energy development. Since 2004, the World Bank's Infrastructure Department of the East Asia and Pacific Region and China State Council Development Research Center's Research Department of Industrial Economy have jointly conducted a study program on China's Sustainable Energy Development. The objective of the study was to provide a comprehensive set of policy recommendations for integrated national sustainable energy development. The recommendations are based on analyzing China's current energy status, referencing international advanced experience and lessons, investigating energy-related industrial reform and development, and evaluating the effectiveness of the current available energy policies.

Through joint efforts, the two-year study program was completed successfully in June 2006. I hope the publication of the study materials will help people in China and the world gain sufficient understanding of China's energy status and facilitate input to China's sustainable development programs. I would like to thank sincerely all the scholars, experts, and my colleagues who have made tremendous contributions to the successful study program.

Qingtai Chen
Standing Member of Chinese People's Political Consultative Conference
and Former Vice President of the Development Research Center of
The State Council, China

Foreword

Nearly 30 years of China's remarkable economic growth have caused and required dramatic changes in many sectors. The nation's energy economy is no exception. It has supported that growth with a generally adequate and relatively low-cost supply of energy, creating in the process the world's largest coal industry, its second-biggest oil market, and an electric power business that is the global leader in terms of annual capacity additions.

But this report points out that while underpinning economic growth, the energy sector has increasingly been stressing the natural environment, placing heavy demands on domestic energy sources, and exposing the country to the risks of dependence on foreign petroleum supplies. If energy requirements continue to march in lockstep with economic growth, doubling every decade, it will be impossible to meet the energy demands of the present without seriously compromising the ability of future Chinese generations to meet their own needs for energy. Briefly, a business-as-usual future appears unsustainable.

Happily, there is a high level of environmental awareness on the part of China's decision makers—reflected, for example, in proposed atmospheric pollution caps and in relatively advanced environmental policies—combined with an obvious preparedness to act in the area of energy law and institutions. This report strongly reinforces these policy directions by demonstrating the need for strong and effective measures to rein in energy demand growth, to integrate energy and environmental policies, and to address security of supply.

Pursuit of a less energy-intensive growth path than was followed by other industrializing countries is a daunting challenge but certainly not beyond China's capabilities. It will require innovative thinking, the development of a comprehensive policy framework, and a coherent implementation strategy. Economic development strategies will have to take account of energy considerations, possibly emphasizing less energy-intensive, higher value added sectors to help achieve the government's objective of dramatically reducing the energy intensity of the economy by the end of the 11th Five-Year Plan (2006–10).

International relations have a critical role in achieving these goals, for example, in improving the security of imported energy supply, in achieving advanced technology transfer, and in cooperating in the area of global climate change. As one of the world's energy giants, affecting and being affected by global climate change, China has a vested interest in addressing this issue. The Clean Development Mechanism under the Kyoto Protocol provides a strong incentive for countries with greenhouse gas reduction commitments to invest in emissions-reducing projects in China. There is hence a large potential to create a revenue stream that can be used both to reward Chinese emissions reducers and to fund the technological leapfrogging that can be critical in achieving further energy efficiencies and environmental improvements.

The World Bank was pleased to cooperate closely with the Development Research Center under the State Council on the present report and stands ready to support the government of China in whatever follow-up to this report it may require.

David Dollar
China Country Director
East Asia and Pacific Region
The World Bank

Foreword

The World Bank has for more than 15 years supported the efforts of the Chinese government to develop and revitalize the country's energy sector. This has been a time of rapid energy growth, pacing and supporting the national economy; of significant structural change, allowing market forces to secure efficient resource use; and of corporate renewal, with state oil and power enterprises emerging as listed companies on the international stage.

The Bank's advisory work has covered a large range of energy activities with a particular focus on the electric power subsector and renewable energies and, to a lesser extent, on coal. In 2001, a joint report addressed structure, reform, and regulation in the oil and gas subsector, and much cooperative work followed on regulatory and legal issues in the downstream natural gas industry; the objective was to initiate implementation of a modern regulatory structure to ensure a sustained development.

In late 2003, it was decided with the support of the Development Research Center (DRC) under the State Council to step back; take a broad view of where China's energy economy has come from and where it is now headed; review the implications for the sector and for broader environmental, security, and societal concerns; and propose policy responses for consideration at the highest level. This joint effort of the

DRC and the Infrastructure Department of the East Asia and Pacific Region of the World Bank attempts to address the sustainability concerns that emerged from the unbridled growth of energy consumption during the 10th Five-Year Plan (2001–05).

This report is not a roadmap toward sustainability. To attempt such a task would be both presumptuous and beyond the resources available for the activity. Instead, the report summarizes areas of policy emphasis developed during and before this cooperative task, some of which have already been reflected in policy announcements made by high-level decision makers. It proposes a set of building blocks for an urgently needed, new, comprehensive energy policy to prepare a blueprint to put the energy sector on a sustainable path.

Christian Delvoie
Director
Infrastructure Department
East Asia and Pacific Region
The World Bank

Preface

As an emerging global giant, China is playing an increasing international energy role, trading extensively, affecting and being affected by world markets, investing abroad and receiving foreign investment in its domestic sector and cooperating in bilateral and multilateral contexts. The nation's large and growing emissions of greenhouse gases are attracting much attention in the context of the global climate change issue. They are worrisome in aggregate terms and for their potential growth, even if they are at the same time still relatively small on a per capita basis.

China's energy sector must now rapidly adapt to function effectively in a new global energy context characterized by oil prices that are high by historical standards, competition for resources, rising concerns about supply security and a strong focus on climate change associated with energy use. The sector also faces internal challenges as unchecked consumption growth during the 10th Five-Year Plan (2001–05) poses huge demands on the energy-supplying industries and services and threatens the sustainable development of the sector.

Given the current stage of energy technology advances, China does not have to follow a steeply rising curve of energy intensity to reach high levels of personal income, as did the advanced industrial countries. Instead, the nation could develop along a new path that would improve the economic well-being of its people while avoiding the adverse impacts of rising energy intensity on the economy and the environment.[1]

This is the path of energy sustainability. It will enable China to meet the needs of the present without compromising the needs of future generations. The road toward sustainability is not an easy one. On the domestic front, changes must be made and practical steps taken to rein in energy consumption, properly integrate energy and environmental policy, improve the security of energy supply, and get correct price signals to energy producers and consumers to ensure adequate and reliable supplies of environmentally and socially acceptable forms of energy at competitive prices. On the international front, China should be engaged and assisted in meeting the daunting challenges ahead rather than vilified and blamed for all the energy-environmental sins of the world. Access to the most advanced technologies through increased and mutually beneficial international cooperation is a prerequisite to success, but the initiatives already taken in this field are dwarfed by the momentous challenges ahead. More alarming, the window of opportunity to make these changes and take the required policy measures and cooperative steps is closing rapidly because of the huge current growth supported by investments in established rather than advanced energy technologies.

This report proposes the development of a coordinated and comprehensive national policy within the "Energy Law" that is presently under preparation based on four foundational themes: reducing energy growth below economic growth, making better use of national energy resources, safeguarding the environment, and making the energy system robust to withstand potential disruptions. It stresses that the policy measures and program to achieve sustainability will have to be the subject of careful consultation within government and extend to mobilizing China's civil society. The scope of and potential for fruitful international cooperation is also explored, but the detail must be worked out in a cooperative framework. In that connection, the current international focus on global warming will likely give rise to forums in which some issues, such as international technology exchanges, can be worked out. Support for this view has just come from a major new report, related to global warming, which states that "coherent, urgent and broadly-based action on energy research and development requires international understanding and cooperation embodied in a range of formal multilateral and informal arrangements." It claims that cost reductions can be boosted "by increasing the scale of new markets across borders." Scale economies are one of the things that China is uniquely positioned to offer other countries.[2]

The main body of the report is arranged in six chapters. The first examines what the projections say about the future of China's energy consumption and concludes that urgent action is needed to avoid locking the country into an unsustainable energy development path. Chapter two evaluates end-use efficiency and finds that a less energy-intensive path can be founded on the most advanced technologies. The third chapter examines the damaging environmental impacts of the huge prospective energy growth, which dictate the need for a less-intensive path, a larger share of clean energy sources, and dramatically more clean coal. Chapter four assesses security of energy supply and proposes means to improve it and ensure the safety of national energy supply sources and networks. The fifth chapter emphasizes getting right the pricing fundamentals of the sector. The last chapter draws conclusions from the first five on matters requiring urgent policy attention and proposes development of a coordinated and comprehensive national policy for energy sustainability.

Notes

1. Berrah, Noureddine. 1983. "Energie et Développement." *Revue de l'Energie*, Special Issue, August/September.
2. Review of the Economics of Climate Change. A Report by Sir Nicholas Stern for the Government of the United Kingdom, Part V *International Collective Action*, p. 516. Web-based pre-publication version, October 2006.

About the Authors

Noureddine Berrah is an energy economist who spent 15 years in the Société Nationale d'Electricité et du Gaz (SONELGAZ), an Algerian power and gas utility, before leaving the company as Deputy Director General of Development in 1985. After three years as a Directior General of the Entreprise Nationale des Systemes Informatiques, he joined the World Bank in 1987, where he worked on several energy projects and technical assistance activities related to power and gas sectors restructuring and regulation with special focus on China. He continues to provide consultation on energy issues in China, Indonesia, and the Maghreb countries.

Fei Feng is Director of the Industrial Economics Research Department and Research Fellow at the Development Research Center (DRC) of the State Council of China. He began his work at DRC in 1993. His research focuses on industrial development policies and restructuring and regulation of monopoly industries. He participated in many research projects at the DRC, including *National Energy Comprehensive Strategy and Policy of China by 2020, Strategic Restructuring of China's Economic Structure, WTO's Impact on Auto Industry, Reforms and Restructure of the Monopoly Industry, Strategies for Sustainable Development of the Power Industry in*

China, and *The Regulation Reform in Power Industry in China*. He has received numerous awards from the Chinese government.

He received his Ph.D. in 1991; from 1991 to 1993, he pursued postdoctoral studies at Tsinghua University. In 1994, he received special training in public policy at Carleton University and the University of Toronto in Canada. He is also a senior consultant to several government departments.

Roland Priddle is an economist who spent 33 years in the government of Canada, latterly as Chairman of the National Energy Board. Since his retirement nine years ago, he has been a consultant for several energy studies by the World Bank in cooperation with Chinese institutions. His principal professional interest is in issues of energy policy and regulation.

Leiping Wang is an energy specialist who spent 10 years with the Beijing Economic Research Institute of the Ministry of Electric Power in China and four years with a global consulting firm based in London and Hong Kong (China). He joined the World Bank in early 2004, where he works on energy projects in China, Indonesia, and Timor Leste.

Acknowledgments

This report was derived from review of available (but not always published) Chinese and English studies and reports. It draws on the ideas, knowledge, and experience of a diverse group of Chinese and international experts. The analyses and consultation have been carried out by a joint task team of the Infrastructure Department of the East Asia and Pacific Region of the World Bank and the Research Department of Industrial Economy at the Development Research Center (DRC) under the State Council.

Composition of the Joint Task Team

Group leaders: *Noureddine Berrah*, task team leader, Energy and Mining Sector, World Bank
Fei Feng, director-general, Research Department of Industrial Economy, Development Research Center

Members:

World Bank Team
Roland Priddle, consultant
Leiping Wang, senior energy specialist
Ximing Peng, energy specialist
Peter Miles, consultant
Matthew Mitchell, consultant
Kazim Saeed, energy specialist
Jaspal Singh, consultant
Yunshi Wang, consultant
Cristina Hernandez, team assistant
Mary Fisk, production editor
Pat Katayama, acquisitions editor

DRC Team
Shi Yaodong
Liang Yangchun
Lai Youwei
Wang Jinzhao

Wang Qingyi, energy statistician and consultant to the DRC prepared or reviewed and adjusted all energy statistics used in the report to meet accepted international standards and to ensure the cohesiveness and quality of the data used in the analyses.

World Bank staff members who served as peer reviewers included Kari J. Nyman, lead specialist, Infrastructure and Energy Service Department, Europe and Central Asia Region; Bent R. Svensson, program manager, Policy Division, Oil, Gas, Mining and Chemicals Department; and Zmarak Shalizi, senior research manager, Development Research Group. Jean-Marie Chevalier, Université de Paris—Dauphine; Athar Hussain, London School of Economics; and David Moskovitz, Regulatory Assistance Project, United States, served as external expert reviewers. Comments and contributions were provided by Sally Hunt of NERA Economic Consulting and Roberto D'Addario of Mercados Energeticos during the preparation of the chapter and appendixes related to power pricing. The joint task team is grateful for their insightful comments and suggestions, which shaped the final version of the report.

The team also greatly benefited from comments provided by institutional reviewers, including Bert Hofman, lead economist–China and chief, Economics Unit, World Bank Office: Beijing (China/Mongolia); Gary Stuggins, lead energy economist, Energy and Water Department Energy Unit; and Apurva Sanghi, senior economist, East Asia Infrastructure Sector department, also provided valuable comments on chapters. The team also appreciated the guidance and support provided by the management of the World Bank and the Energy Sector Management Assistance Program (ESMAP), including David Dollar, China country director, East Asia and Pacific Region; Christian Delvoie, director for infrastructure, East Asia and Pacific Region; Junhui Wu, sector manager, Energy and Mining Sector Unit, East Asia and Pacific Region; Jamal Saghir, director, Energy and Water Unit; and Dominique Lallement, adviser, ESMAP.

The report has been reviewed in a dissemination workshop held in Beijing on June 1 and 2, 2006. Summaries of valuable feedback from the workshop on six important issues are presented in appendix E. The report has been revised to incorporate, to the extent possible, the comments provided during the workshop. The joint task team is indebted to all Chinese decision makers and Chinese experts for their contribution.

The work would not have been possible without financial support from ESMAP to carry out the analyses and research activities and from the World Bank to hold the dissemination workshop and Decision Makers Meeting in Beijing.

Although there was broad consensus on the need for developing a comprehensive and sustainable energy policy, the choices, instruments, and recommendations were not as easy to agree on. The joint task team would like to believe that the opinions and recommendations are shared by all who contributed to this report. However, the team leaders remain solely responsible for all the views expressed, all the conclusions reached, and any errors of fact and interpretation that remain.

Acronyms and Abbreviations

APEC	Asia Pacific Economic Cooperation
ASEAN	Association of Southeast Asian Nations
CDM	Clean Development Mechanism
CER	certified emission reduction
CERM	Coordinated Emergency Response Measures
CNPC	China National Petroleum Corporation
CRESP	China Renewable Energy Scale-Up Program
CSO	civil society organization
DRC	Development Research Center (of China's State Council)
EEIA	Energy Education and Information Agency
EOSS	Emergency Oil Sharing System
ERI	Energy Research Institute
ESMAP	Energy Sector Management Assistance Program (of the World Bank)
EU	European Union
FGD	flue gas desulfurization
gce	grams of coal equivalent
GDP	gross domestic product
IEA	International Energy Agency
IGCC	integrated coal gasification combined cycle

IOC	international oil company
LNG	liquefied natural gas
MARKAL	MARKet ALlocation (computer model)
mtce	million tons of coal equivalent
NBS	National Bureau of Statistics
NDRC	National Development and Reform Commission
NOC	national oil company
OAPEC	Organization of Arab Petroleum Exporting Countries
OECD	Organisation for Economic Co-operation and Development
OPEC	Organization of Petroleum Exporting Countries
R&D	research and development
REPL	Renewable Energy Promotion Law of 2005
RPI	retail price index
SEPA	State Environmental Protection Agency
SERC	State Electricity Regulatory Commission
SOE	state-owned enterprise
tce	tons of coal equivalent
toe	tons of oil equivalent
TSP	total suspended particulates
UGDC	urban gas distribution company
UNFCCC	United Nations Framework Convention on Climate Change
U.S. DOE	United States Department of Energy
WGEST	Working Group on Energy Strategies and Technologies
WTO	World Trade Organization

Executive Summary

China's energy sector has come a long way since the start of the open door policy in 1979. Energy supply has broadly met the needs of rapid economic expansion. Now, the sector includes the world's largest coal industry and its second largest power industry, one that is adding capacity at an unprecedented rate and scale. Since the start of the economic reforms in 1978, China has dramatically cut down on wasteful use of energy, which was common, and its energy intensity is now comparable with or even lower than that of peer economies such as India and the Russian Federation. Environmental concerns are being considered at this relatively early stage of the country's development.

These achievements could not have been accomplished without the reforms that have transformed and commercialized the sector. Pricing reforms merit a special mention: energy prices have been gradually increased to reach—and sometimes surpass—average long-term marginal costs. Also, markets are increasingly allowed to determine prices.

The achievements of the past are impressive in terms of policy and results and, in many ways, unique for a fast-industrializing economy facing a dual transition—from a command-and-control to a strong market orientation and from a rural to an urban society. However, China's energy sector faces serious problems that, if not addressed promptly, could jeopardize these achievements.

The Problems China Faces

The root problems are structural and arise from inadequate (or occasionally completely absent) incentives for efficient use of resources, from lack of integration of the energy and environmental policies, from a timid approach to technological innovation, and from fragmented (and too centralized) policy making, given the size and diversity of the energy sector. Understaffed and inadequately financed, the responsible government institutions lack the capacity to ensure adequate and full implementation of laws and regulations relating to the energy sector and protection of the environment.

The sector faces five major outstanding issues:

1. China's enormous prospective energy needs could pose major issues technically and in relation to resource adequacy.
2. Attempts to meet these needs could result in unacceptable environmental damage.
3. Dependence on oil imports and electricity shortages could seriously disrupt the economy.
4. The disconnect between energy policy needs and regulatory oversight is widening.
5. China's energy and environmental institutions do not meet the country's needs.

China's Enormous Prospective Energy Needs

Already enormous, China's prospective energy needs, if left unchecked, will soon grow to a level that may be difficult to satisfy. The highest energy projections of the late 1990s and early 2000s predicted that China's energy needs would grow from about 1.3 billion to 3.3 billion tons of coal equivalent (tce) in 2020. This estimate assumes an elasticity of energy to gross domestic product (GDP) of 0.5, a value close to or slightly higher than that observed during the 1980s and 1990s.[1] Instead, elasticity rose to about 1.0 during the period between 2000 and 2005, and energy consumption surged to about 2.2 billion tce by 2005. China seems to have embarked on a more energy-intensive path. If elasticity remains close to 1.0, energy consumption in 2020 would grow to more than 5 billion tce, more than 50 percent above the value predicted by the highest established forecasts. Energy use on this gigantic scale will pose major issues both technically (for example, in coal transportation) and in relation to resource adequacy (for example, sustaining large coal output or increasing annual oil imports).

Unacceptable Environmental Damage

Supplying unrestrained needs on this unprecedented scale will result in unacceptable environmental damage. Despite the progress achieved during the past decades, pollution—especially from use of coal and increasingly from motor vehicles—remains a serious threat to the environment. All projections show that coal will still account for 60 percent or more of China's primary energy consumption in 2020. In the absence of profound changes, the fast-growing energy consumption will pose a serious threat to the environment, likely reaching crisis proportions by 2020. In all three scenarios developed by the Development Research Center (DRC) and the Energy Research Institute (ERI), emissions of pollutants (nitrogen oxides, sulfur dioxide, smog, and particulates) will exceed the sustainable limits envisioned by the government. China is also on its way to becoming the world's largest emitter of carbon dioxide, even if its per capita emissions remain low compared with industrial countries.

Growing Oil Import Dependence and Electricity Shortages

Growing oil import dependence and electricity shortages are sources of an increasing sense of insecurity and pose a serious risk of disruption to the economy. Given the present institutional conditions, domestic oil output is expected to peak at about 200 million tons in 2015. In 2020, demand may range between 450 million tons (with substantial demand reduction efforts) and 610 million tons (with ordinary efforts). These amounts imply that at least half and perhaps as much as two-thirds of the demand will have to be met through imports. These levels of dependence on oil imports are not unprecedented: most industrial countries rely on imports to meet an even higher share of their oil consumption. Japan's high growth period was fueled very largely by oil imports, which still meet nearly half that country's energy needs. However, China is a late entrant to the international oil market, which is undergoing important structural changes that most likely will result in higher-than-historical prices.

The record of electricity reliability in China has been relatively good, but power shortages resulting from underinvestment in the late 1990s, in both supply and energy efficiency, have had serious economic consequences and stalled power subsector reforms. These shortages have been acute in parts of the country's high-growth belt and caused sizable losses of industrial output. They have triggered large investments in relatively small, inefficient, and polluting generating capacity, both by power suppliers and by industrial users. Such investments have exacerbated local pollution and undermined moves toward environmental sustainability because these newly built assets will be in operation for decades to come. Total security and reliability are not achievable at any reasonable cost, and measures to address oil insecurity and electricity reliability will come at a price to the economy and society. These measures need therefore to be carefully assessed and gradually implemented to meet reliability standards set by the government.

Widening Disconnect between Energy Policy Needs and Regulatory Oversight

Despite the progress achieved in reforming the sector, there is a widening disconnect between the policy needs of a dramatically changed and increasingly market-oriented energy sector and the slow evolution of command-and-control oversight and regulation. The energy sector is characterized by rapid volume growth, a quickly changing energy mix, and increased interaction with unregulated international markets, mainly for crude oil and oil products but also for natural gas and coal. Energy permeates all economic activities and is essential to sustain the country's high economic growth and the comfort and mobility needs of the population. With economic opening, energy consumption is increasingly conditioned by very decentralized economic and lifestyle choices, while policy is still mostly based on command and control. The reform of energy markets and pricing has stalled, leading to regressive actions such as the capping of oil products prices in 2005, which prevented the flowthrough of international prices to domestic markets, thus eliminating refining profits and damaging the industry's financial position and the financing capability of national companies. Prices of energy commodities are providing the wrong signals to consumers because they do not include the social costs of environmental externalities and because they favor increased supply over efficient-use measures (for example, power-saving investments cannot be recovered through the current pricing system). Power subsector reforms are well under way, but implementation also slowed somewhat during the 10th Five-Year Plan (2001–05) because of emerging vested interests and inadequate pricing, especially of transmission services.

Inadequate Energy and Environmental Institutions

Policy making is piecemeal and uncoordinated, and implementation, supervision, and regulation are weak, mainly because the government's energy and environmental institutions are understaffed and underfunded. By comparison with most industrial and even developing countries, the

Chinese institutions concerned with energy and the environment are inadequate. As a consequence, policy making, supervision, and regulation of the sector do not match the challenges it faces. Implementation of energy policy has suffered from lack of or distorted incentives or failure to integrate energy and environmental goals. For example:

- End-use energy efficiency is not promoted by power companies because they are not allowed to recover efficiency-improving investments in their tariffs, and the reduction of electricity consumption would cut their revenues.
- Contrary to the general policy aim of maximizing use of renewable energy and minimizing air pollution, coal-based power generation within a province is, for tax reasons, preferred to cleaner (even cheaper) hydropower imports from other provinces. The power trade is in most cases still dictated by higher corporate levels or by the government rather than by economic optimization considerations.
- Environmental policies have not been updated to meet the needs of emerging generation markets and to prevent extensive use of older and highly polluting coal plants.

Conclusion: An Urgent Situation

The conclusion is unavoidable: present growth trends are unsustainable; the situation is urgent and calls for new and coordinated policy making, strengthened institutions, and effective, results-oriented implementation.

China cannot sustain a growth rate of energy demand equal to or faster than the growth rate of its GDP without compromising the ability of future generations to meet their environmental and energy needs.[2] Change is urgently needed because these energy-intensive assets have long lives and China's new stocks of them, from power generation assets to automobiles to buildings, are growing rapidly. Any low-efficiency equipment (for example, conventional coal-fired power plants) or buildings added in the coming years will still be in operation in 2020 and far beyond. Furthermore, international experience during the past three decades shows that energy-use practices alter slowly and that lead times for fundamental changes are likely to be long. Some results can unquestionably be achieved by the end of the 11th Five-Year Plan (2006–10) with special efforts to conserve energy, but sustainability will require a longer-term economic and societal vision. The issues that underlie China's current energy situation and the risks of being locked into an energy-intensive future are recognized by the government. In response, it is developing a comprehensive strategy, to be embodied in a new energy law currently under consideration. To be effective, this important step should be complemented by institutional change and improved policy implementation.

Sustainability: A Challenging but Feasible Goal

The need for change has been recognized at the highest levels of decision making. Most of the suggestions for achieving sustainability discussed during the preparation of this report have been, to some degree, adopted by the government. This report is not a roadmap for achieving sustainability. Rather, it offers an overall strategic direction; it highlights themes that should guide policy making to achieve sustainable development; it identifies institutional and other changes to ensure their adequate implementation; and it highlights a program of action to achieve a 20 percent reduction in energy intensity, which is one of the key objectives of the 11th Five-Year Plan.

The 20 percent reduction in energy intensity called for during the 11th Five-Year Plan is a first short-term step toward sustainability. Nevertheless, it is extremely ambitious and challenging (especially following the upward revision of GDP and energy consumption figures of the past decade). Reducing energy intensity by 20 percent would require a reduction of energy consumption by more than 600 million tons of coal equivalent relative to the business-as-usual scenario.[3] China managed, on average, comparable reductions in energy intensity between 1980 and 2000, but the potential for easy savings has been mostly tapped. Future reductions will require better-targeted incentives and improved policy measures.

The 20 percent goal is feasible for three reasons. First, there are still significant efficiencies to be achieved in energy-intensive sectors such as ferrous metals, where state-owned enterprises (SOEs), which are subject to direct government influence, are still the major players. Second, the present environment of high energy prices (which is expected to continue), if passed through to consumers, strongly encourages the search for efficiencies. Third, an active program that offers additional incentives where justified is contemplated. However, sustainability will require a longer-term vision and enduring policies that are not yet in place.

The Four Pillars of Energy Sustainability

Four pillars should guide policy making for energy sustainability—less energy-intensive economic growth, better use of domestic resources, safeguarding of the environment, and enhancement of supply security.

Less energy-intensive economic growth. Energy consumption growth needs to be kept lower than economic growth to the greatest extent possible. The growth trajectory of energy must be brought well below that of the economy: the DRC and ERI's projections indicate that sustainability requires an elasticity lower than 0.5, a daunting challenge for China because such a level has never been achieved by any industrial country during the rapid industrialization phase. Achieving it will require dramatic improvements in the efficiency of energy conversion and consumption across all economic sectors and all areas of personal energy use. The pace of technological change in the energy sector must be accelerated, and innovative policies to encourage energy efficiency must be adopted in every sector. Special focus is needed on three points:

- Tapping the sizable potential of energy efficiency in the existing capital stock at all stages of the energy cycle
- Embodying best-in-class technology in the new stock of equipment for energy production (particularly coal and oil); energy conversion (thermal electricity); transportation (rail, pipeline, and cars and trucks); and energy use (motors, lighting, heating, cooling, household appliances, and industrial processes)
- Bringing about lifestyle changes by promoting novel urbanization models, with special focus on advancing the development of compact, medium-size cities and enhanced public transportation to meet the population's mobility needs.

China can achieve the first two goals in part by leapfrogging over the intermediate stages of technical and structural development that have been followed by other industrial countries. The potential for obtaining and deploying cutting-edge technologies in this way can be enhanced by exploiting the possibilities offered by international energy cooperation, particularly in the context of the Kyoto Protocol and the Clean Development Mechanism (CDM).

Better use of domestic resources. The country's sizable energy resources need to be better harnessed and used. Leaving aside coal, which will continue to dominate the energy sector in the foreseeable future, oil and gas resources need to be harnessed in a more systematic and market-based way. China's landmass, including its offshore areas, undoubtedly contains large resources—some of which are underused and others that are as yet undiscovered.[4] Maximum development of those resources, on an internationally competitive basis, will help minimize energy import needs and related security and financial concerns. Under the right policy conditions, renewable energy forms can undoubtedly meet a growing share of requirements, and there is a place for nuclear energy as well, if long-term safety and security issues are properly addressed.

Safeguarding of the environment. The environment must be safeguarded from the adverse effects of energy production, conversion, and consumption at the national and local levels. Environmental degradation caused by the present energy system, particularly by conventional methods of coal use, is at an unacceptable level and poses a serious hazard to human health. The challenge is to avoid the "pollute during the industrialization phase and clean up later" model followed by most countries. Meeting this challenge will require increased focus on end-use energy efficiency, low-emission or low-carbon technologies such as renewable sources, nuclear energy (with adequate security and safety measures), and, most importantly, clean-coal technologies—preparing, wherever possible, for carbon dioxide sequestration.

Enhancement of supply security. The energy system must be better prepared to withstand supply disruptions, however caused, and vital energy installations should be made more secure. China's inevitable and accelerating integration into the global hydrocarbons market creates the opportunity to diversify its energy balance and build into it more "clean" natural gas. At the same time, this integration poses the challenge of addressing concerns about security of supply, particularly of oil imports, and about vulnerabilities of critical energy infrastructure such as liquefied natural gas import terminals. Enhancing security of energy supply calls for a suite of measures ranging from embodying the right fundamentals in a sound national energy policy to better overall energy planning, to taking highly specific steps such as establishing emergency oil stockpiling and allocation schemes. The scope for greater contribution from greener renewable energy forms (hydroelectric, with required environmental and social safeguards, and wind, solar, and biomass) should also be emphasized. Because gas and electricity will account for an ever-growing share of national consumption, energy infrastructure will become more vulnerable. Special attention needs to be given to enhancing the security of energy networks and supply sources. Although coal must and will remain the mainstay of China's energy supply, diversification of the mix of energy forms, sources, and suppliers can generate important environmental and security benefits.

A Shift in Policy Making

More fundamentally, however, sustainability requires a paradigm shift in policy making. Pursuing a less energy-intensive development path than was followed by other industrial countries is a tremendous challenge but is not beyond China's capabilities. China could emulate the approach taken by Japan during the reconstruction period following World War II. Japan took advantage of the most advanced technologies available at the time to become one of the most energy-efficient and dominant trading economies in the world.

International experience indicates that development of energy management services and renewable and energy-efficiency technologies spurs growth and creates jobs. Development of the most advanced energy technologies could change China from a laggard to a global leader, leapfrogging some stages of development to boldly deploy technologies that are slowed in more advanced countries because markets there are not large enough, stranded costs are high, and opposition by vested interests is too strong.

Energy markets are undergoing tremendous evolution, in part because of the current high level of prices and increased awareness about climate change. China could be at the forefront of technical change, given the size and huge growth of its energy sector. In no other country have both the required market opportunity and the political will to embark on such an undertaking coincided. Now, there is a clear indication that both these requirements can be met by modern China. Development of a coherent strategy to implement a less energy-intensive path will require rethinking and adjusting important elements of the nation's economic and societal development strategy. It will necessitate fundamental lifestyle choices that will call for a national consensus and the continuing support of civil society. It will also require a shared strong policy commitment combined with consistent and effective implementation.

The Path to Sustainability

Sustainability is a long-term objective that requires a shared vision and a progressive approach, since "China seeks its objectives by careful studies, patience, and the accumulation of nuances," observed Henry A. Kissinger, a former U.S. secretary of state (Kissinger 2005, A19). However, it is important not to conceal the growing need to avoid locking the country into an energy-intensive growth path and ways of life that are founded on high personal energy consumption, as experienced by more industrial countries.

Developing an Integrated Energy and Environmental Policy

The energy sector is unlikely to be put firmly on a sustainable path without the adoption of innovative and coordinated policies grounded in a sound legal framework. The energy law currently under consideration provides a timely opportunity to develop a comprehensive legal vehicle to support the formulation and implementation of an integrated, coordinated, effective, and environmentally sound energy policy. The law must have as its main guiding principles the four policy pillars. It must also achieve the following:

- Direct vital national lifestyle choices toward a resource-conscious society.
- Be flexible enough to accommodate and encourage novel approaches and initiatives as they develop and adapt them to the diverse local conditions.
- Be backed by a comprehensive set of focused and policy-driven studies in order to chart a course of specific actions with measurable outcomes, to put the energy sector on a sustainable path.

The recently established Energy Commission under the State Council should take stock of the large number of existing studies, identify any further analyses needed, and initiate them as soon as possible in cooperation with all concerned agencies. The focus should be on themes that are vital to sustainable development at the national level:

- *Pricing and taxation issues.* In particular, the focus should be on the progressive incorporation of external environmental costs in energy pricing and on incentives for efficient energy use.
- *Technology choices.* Particular attention should be paid to leapfrogging to clean-coal and other low-emission and climate change–friendly technologies, including strategies that consider using low-cost CDM transactions to acquire the technologies and for-cost reductions through adaptations to local conditions to deploy them. Novel approaches to urbanization and personal transportation are also needed. Intersectoral planning and coordination is a must, considering that two major future drivers of energy consumption are urbanization and road transportation.
- *Enhanced international cooperation.* Through mutual efforts, China and the international community can work together to support technology transfer efforts and further energy security based on mutual benefits.

Implementing the New Framework

When the central government, with appropriate regional consultation, has developed the energy law, consideration should be given to delegating to the provinces and lower administrative entities the detailed working out of the law to achieve predetermined national objectives. Under this scenario, the provinces and other entities would become operationally responsible for implementing the sustainability policy and would be expected to monitor and report the results. In turn, they could reasonably expect to receive from the central government technical and financial support, where justified, to help them achieve those national objectives.

A strong sense of urgency at all levels of government and society is a necessary premise of the energy sustainability policy, because the window of opportunity to implement changes is closing quickly owing to China's very rapid growth. Furthermore, this tremendous growth will breed vested interests and opposition to change at later stages. Urgency also arises because the costs of the present energy development path are so high in terms of harm to human health, environmental degradation, and energy insecurity. These costs are not reckoned into pricing and GDP accounts.[5] The volume and type of energy investments made in the next few years will determine whether China's energy future will be sustainable. Time is of the essence because, as stressed in a Chinese saying, "If you let the opportunity slip, it may never come again." Consideration must be given to taking strong policy action in the first year of the 11th Five-Year Plan.

Paving the Way toward Long-Term Sustainability

Following are recommendations for consideration by Chinese authorities to pave the way toward long-term sustainability. They relate to a foundational level of policy making. The main report identifies some second-level policy measures, such as oil stockpiling as a means to address energy supply insecurity and particular technologies to achieve more efficient and environmentally friendly energy supply and consumption. Those measures—and many more beside—will need further assessment and evaluation as part of the work that goes into preparing the energy law.

Eight building blocks can take the process beyond the initial policy statement:

Building block 1: Develop an integrated policy within a sound legal framework. A comprehensive, coordinated policy directed at energy sustainability should be expressed first in a public commitment made by the highest level of government and then embodied in the energy law. The public commitment would underscore the government's determination to effect change. It would recognize the long-term nature of the project, and it would stress the importance that will be placed on effective implementation involving broad public consultation; close cross-sectoral coordination with other policies concerned with urbanization, agriculture, transportation, and international trade; and a significant degree of delegation to the provinces and lower-level administrative entities within an appropriate framework determined by national laws and regulations.

Development of the energy law does not have to wait for the completion of comprehensive programs of studies. What is needed is a framework law that will clearly define the central government's purpose in developing and implementing a comprehensive policy for energy sustainability. It should empower the existing institutions and create new ones to complete the needed organizational framework (agencies with clearly defined responsibilities and adequate resources) and give force to the policy that will put the energy sector on a sustainable track and meet the daunting challenges of becoming a resource-conscious society. When technical studies have been completed, consultations carried out, and specific policy implementation decisions made, complementary regulations can be made under the authority of the law, providing for the implementation steps and schedule and ensuring that adequate funding is available for the needed programs.

Building block 2: Focus strongly on efficiency and sustainability and consult the public on these issues. The purpose of this task is to ensure that efficiency and sustainability are maintained as national priorities. While this goal is a well-accepted theme, stakeholders do not always link it to the required behavioral and lifestyle choices that are needed to achieve sustainability. The transition to sustainability cannot be accomplished without important societal changes. These changes require the commitment and cooperation of all stakeholders. Officials, energy producers, industrial and commercial consumers, and private citizens must all be convinced that efficiency and sustainability are fundamentally important energy objectives that will be unwaveringly pursued.

A strong, effective, and informative communications program therefore needs to be developed and put to work by the government to stress the strategic and vital nature of energy efficiency in underpinning the concept of making China a resource-conscious society. The goal of a high degree of public awareness and support would be reached by highlighting the dangers of the present energy path and the advantages—in terms of personal well-being, environmental enhancement, and international competitiveness—of pursuing energy-efficiency and sustainability goals.

The 11th Five-Year Plan's objective of reducing energy intensity by 20 percent is a critical priority. It can be achieved only if drastic and carefully targeted measures are devised and rapidly implemented. Before the end of the first year of the plan, baselines will have to be agreed on, target reductions established by sector, energy-intensity improvements ranked by economic merit, and an action program put in place to achieve at least the 20 percent reduction by 2010. Where economically justified, government incentives such as capital grants, value added tax exemptions or reductions, and other incentives may have to be offered.

Building block 3: Strengthen the institutions of energy policy. The establishment of the Energy Commission and the State Electricity Regulatory Commission (SERC) and the steady progress that is being made toward creating a modern regulatory structure for the downstream gas industry indicate the government's commitment to improved and strengthened energy institutions. Establishing institutions and developing the organizational framework for policy making and for supervising and regulating China's huge and complex energy sector are important undertakings that require time. However, the efforts so far deployed have been partial and limited, leading to inconsistencies among policies that have been developed on a piecemeal basis and to very weak enforcement of laws and regulation.

The first step in reinvigorated efforts is to clearly define the role of the Energy Commission and to properly empower it. The commission should be given a leading role in the preparation of the energy law under consideration and in devising a comprehensive and inclusive consultation process. The commission's work would in part be carried out by interdepartmental task forces under its supervision. The commission should be appropriately staffed, adequately funded to carry out this important mission, and held accountable for its performance.

The second step is to comprehensively review the current government organizational framework and to identify changes needed to improve the coordination, supervision, and regulation of the sector to meet the challenges ahead. It is especially important to ensure that ongoing energy sector reforms and implementation rules are designed to integrate energy-efficiency and environmental goals. This work should be carried out by another task force under the commission's supervision.

In this context, it is strongly recommended that consideration be given to these tasks:

- Reestablish a ministry of energy.
- Properly empower the SERC and clarify its investigative authority and its responsibilities for electricity pricing.
- Create a regulatory body (or an autonomous department within the SERC) for the downstream gas industry.
- Create other government agencies, parallel to or within the Energy Education and Information Agency, especially to promote energy efficiency and technological innovation to achieve sustainability.

Lower administrative levels of government should be involved in the development of implementation strategies. More powers for policy implementation should be delegated to provincial and lower administrative entities, and their responsibility and accountability for results should be increased.

Institutional strengthening will clearly require substantial additional budget allocations for the energy sector. Such allocations will therefore need to be evaluated to ensure that benefits outweigh costs. During the evaluation, it will be important to recall that China's dedicated energy institutions are few in number and loosely coordinated and that enforcement of laws and regulations is deficient compared with countries that have smaller and less complex energy sectors.[6] On that basis, the benefits of institutional strengthening are likely to outweigh the cost incurred.

Building block 4: Get the fundamentals of the sector right. The disconnect between the increasing market orientation of the sector and the slow evolution of its regulation should be addressed as quickly as possible. Market reforms must be completed in the electricity, gas, and oil products industries. The competitive environment sought by the government, requires access by energy generators to eligible customers in the power subsector and more market players in the oil and gas subsectors. Market-based price signals to energy investors and users will help bring about needed changes in the energy capital stock and in usage patterns. They will result in optimal use of all energy resources. Completing the initial reforms and restructuring of the sector to achieve competitive markets are among the policy initiatives required to underpin a new energy policy. Circumstances will undoubtedly arise in which targeted elements of command and control (restrictions on energy use for particular purposes and efficiency standards) will be needed to complement or correct the working of markets and market pricing. Getting the fundamentals right will require, among other things, the following:

- *Completing the unfinished business of energy price reform.* Reforms of energy pricing that have been initiated in the power, oil products, and natural gas subsectors must be completed. Prices formed by the free operation of competitive commodity markets will provide correct signals to energy sellers, buyers, consumers, and investors, thereby reducing waste and misallocation. To the extent possible, market designs should include demand response options.
- *Completing the unfinished business of instituting an effective structure for regulating prices charged by natural monopolies.* Methods for price formation in the sector's monopolistic segments, such as electricity and gas transmission and distribution, should be adopted as soon as possible. Prices and price methods should provide strong incentives to invest in end-use energy efficiency. Such incentives will promote further competition in the markets for the energy commodities transported by these monopolies and provide operators with the revenues required to develop and adequately operate these systems—systems that are vital to reliability of supply.
- *Increasing efforts to incorporate external (mainly environmental) costs and make appropriate use of taxation to address market failures.* This step should begin as soon as practicable. The speed of its implementation must be carefully related to the capacity of the energy sector to adjust. Incremental revenue from higher pollution levies could be used to fund increased efforts to achieve the 11th Five-Year Plan efficiency goals, thereby reducing the net cost to consumers.

Building block 5: Deploy cutting-edge technologies and enhance technology transfer, development, and implementation. The basic theme is to identify international best-in-class technologies at each stage of the energy chain, from production through final consumption of each energy form, to rapidly acquire and develop the technologies best adapted to the country's conditions, and to secure their widest possible application and most rapid deployment. There must be a particularly strong focus on climate-friendly coal technology, especially clean-coal uses with potential carbon dioxide sequestration and various forms of renewable energy. Along with this focus must come a focus on the nuclear technologies envisaged by the government and their required safety and security measures. In several areas, such as urbanization and personal mobility choices, changes in actual or expected social behaviors are necessary to avoid sustaining the recent trend of high energy elasticity, reduce the energy intensity of growth, and ensure security of supply. Achieving these changes will require substantial investments, which should be evaluated on the basis of (a) cost-benefit analyses over the life cycle of the assets and (b) environmental externalities. The substantial revenues from the carbon finance transactions being developed jointly by the government and the World Bank could be used to acquire these technologies, with the dual purpose of achieving China's objectives and contributing to both cost reduction and creation of a market to accelerate the development of these technologies globally.

Building block 6: Broaden international cooperation. The objectives span a broad range of national interests, from technology transfer to emissions trading to improvement in the security of imported energy supplies by means of bilateral and multilateral arrangements. China and Chinese entities are already extensively engaged in international activity, but current multilateral and bilateral initiatives are dwarfed by the growth of the energy sector. These activities need to be better coordinated and significantly scaled up to meet the momentous challenges of energy sustainability and security and to have significant effects on climate change. China should take full advantage of the energy development opportunities presented by such institutions as the Kyoto Protocol and the Asia-Pacific Clean Development and Climate Change Partnership to step up efforts for technology acquisition and deployment.

Building block 7: Mitigate the impacts of energy reforms on vulnerable groups sensitively and with targeted measures. Achieving energy sustainability will entail sweeping policy changes and reforms. Some of these changes and reforms will have specific social impacts on poor communities in isolated locations. For example, the reform of power and gas pricing, including putting transmission and distribution utilities on a sound financial footing, may have adverse effects on low-income customers. Also, rationalization of energy production (closing of small, unsafe coal mines) and conversion (rationalization of the oil-refining industry) may have significant local economic effects.

A necessary precondition for effective policy implementation is therefore to anticipate and prepare for potential adverse effects and to embody in relevant policies and programs special measures to address those effects, so as to minimize their impacts. As a general principle, it is more effective and cheaper in budgetary terms to target those measures at the affected group or locality rather than to provide price subsidies to all consumers of a particular product; in short, subsidize the needy rather than the commodity. For example, if low-income consumers are going to be affected by reform of gas pricing and distribution, it makes better sense to provide specific offsets such as direct income assistance or lifeline rates for poor people than to delay or distort comprehensive pricing reform. As for all policy decisions, proper evaluation and assessment are required. Funding of development initiatives in poor communities affected by the closing of small and heavily polluting energy assets could be secured through the trading of emission reductions, following consultation with the affected communities.

Building block 8: Mobilize China's civil society and social organizations in pursuit of agreed national policy objectives. International experiences after the first oil crises show that such programs have been successful in reducing energy waste and improving efficiency. They have been most successful in countries where dedicated government agencies have been created and are well staffed and funded. Given the importance of energy issues in China and the vital nature of efficient energy use to sustain the competitiveness of the economy and preserve the environment, establishing an Energy Education and Information Agency at the national level is recommended. Such an agency should have a clear mandate to coordinate all activities to make energy efficiency and sustainability a national priority supported by all stakeholders, with a special focus on civil society. The mobilization of society in the cause of sustainability can also be assisted by involving civil society organizations (CSOs, or *social organizations* in China). CSOs can play a significant role in delivering services and implementing development programs, especially in sectors and regions where the government's presence may be weak.

Sequencing the Steps to Sustainability

The government should first vigorously reiterate its strong commitment to an integrated energy policy set in a sound legal framework that is directed at the overarching goal of sustainability. This commitment, expressed in a policy statement from the highest level, should be made as soon as possible to develop a shared vision among all institutions concerned. The window of opportunity is closing; all sectors of the economy and society should be made aware that concrete steps are going to be taken to promptly address energy and energy-related environmental issues.

In the short term, parallel actions are needed both to achieve the 20 percent reduction in energy intensity during the 11th Five-Year Plan and to set up the policy and institutional framework to put the sector on a sustainable track. Achievement of this highly ambitious reduction in energy intensity will require comprehensive and methodical planning and implementation monitoring, involving these tasks:

- Resolve the statistical debate that is casting doubts on the sector's achievements, rethink and overhaul the statistical system, and agree on baselines by sector and on continuous monitoring of the outcomes.
- Establish target reductions by sector, focusing on highly energy-intensive ones and targeting large SOEs.
- Identify best practices in efficiency and demand-side management programs in the power subsector and remove barriers to investment to ensure increased saving in the near term.
- Rank energy-saving potentials by industry, business, and cost.
- Develop a communications strategy for building consensus on the program and supporting its objectives.
- Design and implement, where needed, a program of targeted and performance-based incentives for investments and improved operating practices to achieve the 20 percent reduction.

All these steps need to be taken before the end of 2006 if this ambitious program is to succeed by the end of 2010.

Shaping a future of sustainability will require other tasks:

- Staff, budget, and mandate the Energy Commission to (a) advise on and prepare the new energy law; (b) consult the public, using a well-designed and well-targeted communications program; (c) secure close cooperation and coordination with other government departments; and (d) supervise a program of energy studies carried out by interdepartmental task forces staffed by the nation's leading energy experts from government, industry, and academia.
- Review and remove, as soon as practical, the existing price controls on energy commodities to prevent giving distorted signals to consumers and to encourage responsible use of energy, while safeguarding the poor through targeted and/or performance-based subsidies.

- Consider establishing a ministry of energy, with special responsibilities for policy making and responsibility and accountability for implementation.

In the medium term, but with some work to be initiated as soon as possible,

- The role and responsibilities of the SERC should be clarified and the SERC empowered to work toward having a functioning electricity market in which prices and investment decisions incorporate, as quickly as feasible, environmental externalities.
- A gas regulatory commission should be created with responsibilities for the gas market downstream of the gas purification plants similar to those that the SERC has for the electricity market.[7] As a result, natural monopolies in the gas and electricity industries (the transmission and distribution functions) would be subject to modern regulation.
- Commodity markets for electricity, gas, and oil products should be opened to foster competition and monitoring (rather than control and regulation) of prices, to prevent possible abuse of dominant market positions. As a result, energy commodity markets would be regulated by competition rather than by the government, with consequent efficiencies in the use of all resources and better balancing of supply and demand in all circumstances.
- A program of technology transfer, development, and rapid implementation should be launched, resulting in best-in-class technologies being adapted to and built into China's energy economy, with consequent efficiency and environmental and industrial benefits.
- International cooperation, bilateral and multilateral, should be significantly scaled up to meet the tremendous environmental issues facing the country and the world.
- Improvement in the diversity, flexibility, and security of energy supply should be a continuing, overarching objective.
- The government should respond sensitively to social and local issues stemming from energy reform.
- China's CSOs should be harnessed to the task of consulting, informing, and reacting to energy policy proposals and programs and to the task of implementing policies for sustainability where they can make an effective contribution, particularly in rural and remote areas.

In the long term, beyond the 11th Five-Year Plan,

- The Energy Commission should draw up a program of continuous improvement in energy and environmental efficiency and sustainability.

- The energy law should be updated as necessary to ensure its relevance and to see that the government has all the powers needed to continue on the sustainability path.
- Exploration and development of energy resources should be exposed to investment and entrepreneurship without significant restriction.
- The government should play mainly a monitoring role in energy commodity markets and should intervene only in case of abuses of dominant positions by sellers or buyers.
- The application of internationally best-available technologies internationally should be mandatory in all major new energy production, conversion, and consumption investments and for vehicles and appliances for sale in the domestic market.

Throughout, policy implementation should be characterized by a strong sense of urgency, motivated by the fact that the window of opportunity to achieve profound technical and social changes is closing rapidly.

Table ES.1 (see pp. lii–liii) summarizes the analysis of fundamentals, the steps on the path to sustainability, and the sequencing of those steps and briefly indicates the expectable outcomes.

Notes

1. Elasticity of energy demand to economic growth is measured by the percentage change over time in the dependent variable (energy demand) divided by the percentage change in the independent variable (GDP).
2. Sustainable development "meets the needs of the present without compromising the ability of future generations to meet their own needs" (United Nations Commission on Environment and Development 1987).
3. This reduction amount is greater than the 2004 primary energy consumption of every other country of the world, except Japan, Russia, and the United States.
4. Regarding petroleum resources, in addition to conventional oil and gas (onshore and offshore), undiscovered resources may include oil sands and shales, tight gas, shale gas, and coal-bed methane.
5. According to a joint report published by the Guangzhou Institute of Geochemistry and the Guangdong Institute of Environmental and Soil Sciences, the GDP of Guangdong province would have been

27 percent lower in 2003 if adjusted for environmental costs. One of the leading researchers on the project commented that the ratio of "green GDP" to "conventional GDP" for Guangdong was similar to those calculated by researchers working in other provinces and cities (*People's Daily* 2005).
6. In Canada, at the federal level alone, more than 200 people work on energy policies and programs at the Department of Natural Resources, and about 300 are employed by the regulatory agency, the National Energy Board. The combined personnel budgets are about US$64 million annually. The U.S. Department of Energy's responsibilities are diverse and include nuclear weapons. Its "energy strategic goal" is "to protect our national and economic security by promoting a diverse supply and delivery of reliable, affordable, and environmentally sound energy." The amount budgeted for this purpose in fiscal year 2005 was about US$2.8 billion. The California Public Utilities Commission, the largest state regulatory body in the United States, is responsible mainly for regulating the gas and electricity industries within the state. It employs about 870 people, and its operating budget in 2005/06 was about US$109 million. In Kazakhstan, the Agency for the Regulation of Natural Monopolies and Protection of Competition has a staff of 185 to deal with an open access electricity market of about 70 terawatts per year. France's electricity regulatory commission had a personnel budget for 2005 of about 120 people, and its operations cost about US$30 million annually.
7. A gas regulatory commission could be created either outside the SERC or possibly within it, but autonomous from the SERC's electricity roles.

References

Kissinger, Henry A. 2005. "China: Containment Won't Work." *Washington Post*, June 13.

People's Daily. 2005. "Guangdong Pioneers New GDP Model." July 20. http://english.people.com.cn/200507/20/eng20050720_197250.html.

United Nations Commission on Environment and Development. 1987. *Our Common Future.* Geneva: United Nations Commission on Environment and Development.

Table ES.1 Shaping the Future: The Issues, Guiding Pillars, Building Blocks, and Sequencing Steps to Sustainability

Fundamentals: trends toward an unsustainable future →
Conclusion: present trends unsustainable

Five outstanding issues:
- Scale of needs
- Scale of environmental damage
- Import insecurity and electricity reliability
- Policy disconnects
- Policy implementation

Sustainability: challenging but feasible → Conclusion: paradigm shift needed

Four pillars to guide sustainability:
- Create energy growth lower than economy growth
- Safeguard environment
- Harness resources and prepare for a smooth integration in international energy markets
- Prepare to withstand disruptions

Path to sustainability → Conclusion: initiate integrated, novel, and coordinated policies

Start with a policy commitment; then pursue recommendations for eight building blocks to pave way to sustainability:
- Create an integrated policy in a sound legal framework including an energy ministry and education and information agency
- Focus strongly on efficiency and immediately on achieving 20 percent reduction in energy intensity during 11th Five-Year Plan
- Strengthen the institutions for sector governance (that is, define role of Energy Commission and empower SERC) and improve energy data flows
- Get the sector fundamentals—markets and prices—right
- Enhance technology transfer and development, and speed up deployment (especially clean coal)
- Broaden international cooperation (in security and technology)
- Mitigate potentially adverse impacts on vulnerable groups with targeted measures
- Mobilize civil society organizations to pursue agreed national objectives

Sequencing steps

Short-term steps:
- Establish government commitment to integrated energy policy
- Mandate Energy Commission (advise on law, consult public, and supervise studies)
- Review price controls, and substitute monitoring
- Carry out program of studies

Short-term steps and steps into term of 11th Five-Year Plan:
- Implement aggressive program to reduce intensity by 20 percent by 2010:
  —Agree on baselines and monitor progress
  —Establish target reductions
  —Rank those reductions
  —Develop communications strategy
  —Initiate program with incentives (all by end of year 1 of 11th Five-Year Plan)

Table ES.1 Shaping the Future: The Issues, Guiding Pillars, Building Blocks, and Sequencing Steps to Sustainability—Continued

Medium-term steps:
- Empower and clarify responsibilities of SERC
- Create downstream gas regulator (regulate monopoly prices and access)
- Complete opening of commodity markets (markets regulate commodity prices, which increasingly include externalities)
- Launch and rapidly implement program of technology transfer
- Intensify international cooperation, bilateral and multilateral
- Sensitively respond to social and local issues stemming from energy reform

Long-term steps:
- Establish program for continuous efficiency improvement
- Review energy law to ensure continued adequacy and relevance
- Open energy resources to all investors
- Limit government activity in commodity markets to monitoring and correction
- Implement mandatory application of best available energy-efficient technologies in new investments

Anticipated outcomes

Short-term outcomes:
- Responsibility center created
  —Achieve horizontal coordination across central government (task forces)
  —Achieve vertical coordination with provinces and public groups (consultation)
- Progress toward functioning markets resumed
- Bank of facts and ideas created to draw on for lawmaking

Short-term outcomes, into 11th Five-Year Plan:
- 20 percent established as rallying point and core program created to delink growth in energy and economy, improve environment, and enhance security
- Model for subsequent programs (such as technology transfer) created in terms of analysis, targeting, delegation of implementation, and provision of technical and financial support

Medium-term outcomes:
- Market efficiencies achieved in power and gas subsectors
- Rapid gains in energy technology achieved across all sectors that delink growth of energy and economy
- New policies created on urbanization and transportation that delink energy and income growth
- Diversification increased
- Maximum advantages from international cooperation achieved by levering national strengths

Long-term outcomes:
- Energy law reviewed to confirm permanence of sustainability commitment
- Cleaner, more secure energy mix achieved
- Domestic energy resource development enhanced
- Continuous improvement achieved, best available technologies applied to major new investments, and existing plant updated for further efficiency and environmental gains.

Source: Authors.

CHAPTER 1

Introduction

An Impressive Foundation of Past Achievements

China's energy sector has come a long way since the start of the open door policy in 1979 and the launching of far-reaching economic reforms. Despite periodic electricity shortages, the energy supply has broadly met the needs of the country's extremely rapid economic expansion. Now the sector includes the world's largest coal industry and its second largest power subsector, which is adding capacity at an unprecedented rate and scale. Since the start of the economic reforms in 1978, China has dramatically cut down on wasteful use of energy, which had been common, and its energy intensity is now comparable with or even lower than that of peer economies such as India and the Russian Federation. Environmental awareness is increasing, and environmental degradation has become a major concern among decision makers at the highest level, even though the country's development is at a relatively early stage. Overall dependence on imported energy up to 2005 remained low, at less than 6 percent of total energy supply, because of the massive amounts of cheap domestic coal that dominate the country's energy supply and because of sizable national oil resources.

China's achievements in the energy sector could not have taken place without major shifts in energy policy. These policies have broken state monopolies in the hydrocarbon and power subsectors and initiated the

---

On June 1–2, 2006, a workshop was held in Beijing to disseminate and discuss this report, which was then in late-draft form. The purpose was to obtain feedback from a broad spectrum of Chinese government officials, energy industry representatives, and academics. A number of senior foreign experts also took part. The authors took careful account of the valuable views expressed and incorporated them appropriately in the final text.

development of competitive markets. The government has recognized the fundamental importance of separating its policy and regulatory functions from management and operation of energy enterprises.

In the hydrocarbon subsector, policies have strongly encouraged efficiency and entrepreneurship on the part of the partially listed companies that arose from the former state oil monopoly. Two key results have been the opening of petroleum resource development, at a slow, consistent pace, to foreign investment and the acquisition of advanced technology through production-sharing contracts. Furthermore, as a result of accession to the World Trade Organization (WTO), China is progressively opening its oil products market. The power subsector was transformed in less than two decades from a highly centralized administrative department to a corporatized, commercialized, and relatively open sector.

The government of China has also allowed markets to influence energy prices to a significant degree, gradually permitting them to increase and reach—at times even surpass—the marginal cost of energy supply.

Growing Concerns about China's Energy Future

Despite the impressive achievements in the energy sector, China's policy makers face growing concerns about the country's energy future. From the early 1980s to the mid 1990s, China managed to achieve high economic growth with near self-sufficiency in energy supply. However, during the second half of the 1990s, China gradually emerged as a net importer of energy, and by 2005, its dependence on oil imports had soared to about 45 percent of consumption.

The dramatic escalation of dependence on external oil sources has brought China to the forefront of the international energy scene. As self-reliance in hydrocarbon supply fades into the distant past, coming sharply into focus in China's energy future is a growing dependence on oil imports, probably at increasing prices, combined with burgeoning environmental damage from continued intensive use of coal. Security of supply and sustainable development of the energy sector have become major concerns among decision makers.

The definition of energy security in the international community has changed significantly during the past decade. As box 1.1 indicates, the meaning has evolved from a narrow focus on physical supply concerns to a broader coverage that includes affordable pricing, environmental protection, and social acceptance. Energy sustainability encompasses this broader definition.

Box 1.1

The Evolving Definition of Energy Security

The International Energy Agency

The definition of *energy security* by the International Energy Agency has evolved over the years. For example, in 1985, *Energy Technology Policy* defined it as "an adequate supply of energy at reasonable cost" (IEA 1985, p. 9). By 1993, the definition had grown in complexity, as the following statement of ministers indicates:

> Diversity, efficiency, and flexibility within the energy sector are basic conditions for longer-term energy security: the fuels used within and across sectors and the sources of those fuels should be as diverse as practicable. Energy systems should have the ability to respond promptly and flexibly to energy emergencies. In some cases, this requires collective mechanisms and action. Improved energy efficiency can promote both environmental protection and energy security. Continued research, development, and market deployment of new and improved energy technologies contribute to the objectives. Undistorted energy prices enable markets to work efficiently. Free and open trade and a secure framework for investment contribute to efficient energy markets and energy security. Cooperation among all energy market participants helps to improve information and understanding and helps to promote investment, trade, and confidence necessary to achieve global energy security and environmental objectives (IEA 1993).

The European Union

In its 1990 document, "Security of Supply," the European Commission cited the following definition:

> Security of supply means the ability to ensure that future essential energy needs can be met, both by means of adequate domestic resources worked under economically acceptable conditions or maintained as strategic reserves, and by calling upon accessible and stable external sources supplemented where appropriate, by strategic stocks (European Commission 1990, p. 16).

(continued)

> **Box 1.1 (*continued*)**
>
> By 2004, the definition had broadened to include concerns about pricing: "the availability of energy at all times in various forms, in sufficient quantities, and at reasonable and/or affordable prices [and] ... the availability of oil and gas in sufficient quantities, and in particular on the risk of oil and gas supply disruptions" (Clingendael International Energy Programme 2004, p. 37).
>
> **Academics and Policy Analysts**
>
> The following definitions from academia and policy analysts show a range in focus from physical security to the inclusion of several other aspects, including reasonable pricing, the politics of the Middle East, sustainable development, and the environmental challenge:
>
>> If security of supply is the assurance of the physical availability of oil during a supply disruption, then a country can be said to have achieved this goal if it is always able to guarantee that a given quantity of oil is available with certainty to its domestic market, independently of possible market disturbances (Lacasse and Plourde 1995).
>>
>> ... a condition in which a nation and all, or most, of its citizens and businesses have access to sufficient energy resources at reasonable prices for the foreseeable future, free from serious risk of major disruption of service (Barton and others 2004, p. 1).
>>
>> Energy security has three faces. The first involves limiting vulnerability to disruption, given rising dependence on imported oil from an unstable Middle East. The second, broader face is, over time, the provision of adequate supply for rising demand at reasonable prices—in effect, the reasonably smooth functioning over time of the international energy system. The third face of security is the energy-related environmental challenge to operate within the constraints of "sustainable development" (Martin and others 1996, 4).

For the purpose of this report, *energy sustainability* is defined as access to adequate and reliable supplies of environmentally and socially acceptable forms of energy at competitive prices without compromising the energy needs of future generations.[1] In this definition,

- *Access* denotes the ability to purchase energy from domestic and foreign sources on a commercial basis, with no undue impediments imposed by governments, such as quotas on imports or exports.

- *Adequate* means that energy supplies are sufficient to meet the volume of consumers' needs and will remain so long enough to justify investments in particular kinds of energy processing and in transporting, distributing, and using equipment.
- *Reliable* signifies that the systems for delivering any particular energy form are inherently robust or can become so through technical, commercial, and administrative measures to provide reasonable security against unplanned interruptions.
- *Environmentally and socially acceptable* implies that the production, processing, transportation, and use of energy meet not only current standards of social acceptance and environmental protection but also take into account changes in those standards over time—changes that are likely to make standards more demanding.
- *Competitive prices* refers to better integration in the global energy market to allow China to benefit from access to energy at costs that are comparable with or below those of the country's international trading partners.

The Four Pillars of Energy Sustainability

The core of the analysis in this report concludes that sustainable development of the energy sector would require a national energy policy that embodies the following four pillars:

1. *Ensure efficient use of energy to keep energy consumption growth lower than economic growth to the greatest extent possible.* Industrial countries, even the most efficient ones such as Japan, experienced energy demand/gross domestic product elasticity of 1.0 or more during their industrialization phases. Given the size of China's economy and growth, such high elasticity could lead to unsustainable development of the sector. Shifting to a less energy-intensive path will require dramatic efficiency improvements in energy conversion and consumption across all economic sectors and areas of personal energy use, particularly energy for mobility and household use, two of the most important drivers of growth in energy consumption.
2. *Make optimum use of the country's sizable energy resources.* China's landmass, including offshore areas, undoubtedly contains large energy resources. Some of these are underused, and others have yet to be discovered. Maximum development of these resources, on an internationally competitive basis, will contribute to minimizing energy import needs and related security and financial concerns. Energy

policy, therefore, should ensure the progressive opening of these resources to national and international sources of capital and enterprise, as well as the use of the most advanced technologies for their development. Similarly, they should be fully exposed to international market pricing to ensure that, as in other economic sectors, they are developed on a globally competitive basis.

3. *Safeguard the environment from the adverse effects of energy production, conversion, and consumption.* Serious environmental harm has resulted from China's present energy system, particularly its extensive coal use. Without fundamental policy changes, rapidly rising coal use and accelerating growth in automotive fuel consumption will lead to an intolerable level of environmental degradation. Policy makers must call on all of the available tools to support this pillar, including but not limited to the following: pricing-in of externalities, better and higher taxation of fuel consumption, promotion of advanced and clean coal technologies, maximum economic development of renewable energies, and incentives to channel the universal need for personal mobility toward enhanced public transportation.

4. *Make vital energy installations more secure and be better prepared for supply disruptions.* China's inevitable accelerating integration into the global energy economy has created the opportunity to diversify the country's energy balance and build into it more "clean" natural gas. In the next two decades, this program will be one of the most important. At the same time, this diversification poses the challenge of addressing concerns about security of supply, particularly oil imports, and about the vulnerabilities of critical energy infrastructure, such as liquefied natural gas import terminals and nuclear power stations. Enhancing the security of the energy supply calls for a suite of measures ranging from general pricing and taxation policies to highly specific steps such as emergency oil stockpiling and allocation schemes. In particular, as network energies—notably gas and electricity—account for an ever-growing share of national energy consumption, energy infrastructure will become more vulnerable. Special attention needs to be given to enhancing the security of these vital points.

The Closing Window of Opportunity

In most countries, energy consumption rose in step with or more rapidly than economic growth during their major rapid periods of industrialization. Today, China could avoid an energy-intensive pattern by taking advantage of the most advanced technologies. The advantage arises because most of China's energy-consuming assets have yet to be built: most forecasts indicate that between 2005 and 2020 the number of cars will increase by more than 100 million, electricity-generation capacity will increase by about 600 gigawatts (not including replacement of retired capacity), and building space will increase by about 30 billion square meters. The shift to a sustainable growth path must take place before large capital stocks of energy-consuming goods and energy-intensive lifestyles develop on the European and North American model. Investing in inefficient assets or following energy-intensive lifestyles could lock China into a pattern of energy use that would not be sustainable. The window of opportunity for achieving a sustainable future in this way is closing rapidly because the stock of these assets is growing at an unprecedented rate. Once built, the assets will entail important stranded costs and growing vested interests that would slow future decisions to move toward sustainability. The objective of less energy-intensive economic growth is fundamental. Without it, China cannot achieve security of energy supply and safeguard its environment to achieve sustainable development.

Structure of the Report

This report is a joint effort of the Development Research Center (DRC) of the State Council of China and the World Bank. It is based on a major study by a team of Chinese and international experts, conducted under the leadership of the DRC and the Energy Research Institute (ERI), and on other published sources. It explores possible new approaches to addressing the government's concerns about sustainable development of the energy sector and recommends guidelines for policies to shape the country's energy future along a sustainable path.

Chapter 2 examines what the projections say about China's future energy consumption. These projections are from the late 1990s and early 2000s, covering at least the period from 2000 to 2020. The focus is primarily on the projections of the DRC and the ERI, comparing them with other projections from internationally recognized sources. The report also compares these forecasts with the actual consumption figures for China during 2000 to 2005 and finds that energy demand has grown at a faster pace than envisaged by all studies. The chapter concludes that China has entered a critical phase of economic growth that urgently requires innovative and coordinated policies to avoid locking the country into an unsustainable, energy-intensive development path.

Chapter 3 evaluates the end use of energy by consumers and its efficiency. It reviews China's progress in improving efficiency in energy use and the related policy framework, on the basis of studies by the DRC and ERI, the World Bank, and other institutions, particularly the Energy Foundation. It concludes that following the development path of the advanced industrial countries would not be a sustainable course, given the sheer size of the Chinese population and the fast-growing economy. China can instead devise a less energy-intensive development path that is based on the most advanced technologies available and on new policies that are better adapted to the more decentralized nature of the sector and the changing structure of energy consumption.

Chapter 4 examines the environmental impacts of the momentous growth in energy consumption anticipated during the next 15 years and concludes that, without vigorous policy interventions, environmental pollution and associated degradation will reach unsustainable levels. In addition to following a less energy-intensive development path, China needs to incorporate a larger share of clean energy sources in the energy supply mix and dramatically increase the use of clean coal, which will remain the dominant source of energy for the foreseeable future. Doing so will require substantial investment in the acquisition of the most advanced technologies and in research and development that will enable China to leapfrog to the technological leading edge, especially in clean technologies and carbon dioxide sequestration. This technological leap is necessary if China's economy is to grow and enhance its population's quality of life while preserving and improving the national and global environment. It is also possible, because China has the technical capacity, the market, and the low stranded costs (linked to vested interests) to deploy these technologies.

Chapter 5 discusses China's growing sense of energy insecurity, which is attributable mainly to its increasing dependency on oil and gas imports as well as the country's power shortages during 2001 to 2005. The chapter reviews international experience and makes recommendations on how to develop a market-oriented, multisourced, robust national energy economy to ensure secure energy supplies and the safety of national energy supply sources and networks. Creating such an environment is necessary to minimize disruptions and their negative impacts on the economy.

Chapter 6 emphasizes that one of the most important preconditions for successful energy policy outcomes in an increasingly market-oriented and decentralized economy is getting the fundamentals of the sector right—first, by adequately implementing the energy sector reforms already initiated, and second, by designing a comprehensive, sound energy pricing and taxation policy. Under such a policy, the costs of the environmental damage caused by the different fuels used should be progressively factored into their pricing. Such an approach would provide appropriate signals to consumers to adjust their energy use to take account of environmental effects and to producers to invest in cleaner and more sustainable energy technologies, while meeting economic growth objectives.

Chapter 7 draws conclusions from the previous chapters on urgent matters that need the attention of policy makers. It proposes the development of a coordinated and comprehensive national policy for energy sustainability within the framework of the energy law currently under preparation, based on four foundational themes: (a) bringing energy consumption growth well below the target rate of economic growth; (b) making better use of the country's energy resources, on an economically and environmentally sound basis; (c) safeguarding the environment; and (d) making the energy system more robust and safe so it can withstand supply disruptions. It further advises strengthening the institutions of the sector and improving the less-than-adequate system for collecting and disseminating energy statistics. Finally, the report stresses the need for a constructive international dialogue to give China access to environmentally friendly technologies on a mutually beneficial basis. Such access is essential for successful achievement of the policy goals for energy efficiency and environmental improvement.

Note

1. This definition is adapted from the report of the United Nations Commission on Environment and Development (1987), also known as the Brundtland Report.

References

Barton, Barry, Catherine Redgwell, Anita Rønne, and Donald N. Zillman. 2004. *Energy Security: Managing Risk in a Dynamic Legal and Regulatory Environment*. New York: Oxford University Press.

Clingendael International Energy Programme. 2004. *Study on Energy Supply Security and Geopolitics*. Prepared for the European Commission. The Hague, Netherlands: Clingendael International Energy Programme.

European Commission. 1990. "Security of Supply," *Energy in Europe*, p. 16.

IEA (International Energy Agency). 1985. *Energy Technology Policy*. Paris: IEA.

———. 1993. "Shared Goals." Adopted in a meeting of the IEA ministers, June 4. http://www.iea.org/Textbase/about/sharedgoals.htm.

Lacasse, Chantale, and André Plourde. 1995. "On the Renewal of Concern for the Security of Supply." *Energy Journal* 16 (2): 1–23.

Martin, William Flynn, Ryukichi Imai, Helga Steeg, and the Trilateral Commission. 1996. *Maintaining Energy Security in a Global Context*. New York: Trilateral Commission.

United Nations Commission on Environment and Development. 1987. *Our Common Future*. Geneva: United Nations Commission on Environment and Development.

CHAPTER 2

China's Energy Future
The Challenge of Recent Trends

> **Key Messages**
>
> *Energy consumption trends from 2000 to 2005 indicate that China could be embarking on a faster—and likely unsustainable—energy growth path than earlier major studies of its future energy consumption have suggested.* The Development Research Center (DRC) of China's State Council, in collaboration with the Energy Research Institute (ERI) and with the assistance of international experts, had projected that China's energy consumption would continue growing at a rate substantially less than that of economic growth through 2020. These projections, prepared in 1999 and 2000, reflected the energy and economic growth trends of the 1980s and 1990s. However, actual growth since 2000 has been higher than expected by all studies carried out during the past five years. Between 2000 and 2005, China's total primary energy consumption increased from 1,386 million to 2,225 million tons of coal equivalent (mtce), with energy consumption growing at the same rate as or faster than the economy. As a result, by 2005 energy consumption already had reached 90 percent of the amount in the DRC and ERI's scenario for low energy consumption in 2020 and 67 percent of the scenario for high consumption. The continuation of this trend through 2020 would not be sustainable,

because energy use on this gigantic scale would depend to a considerable extent on massive coal output, which ultimately would face resource and logistical constraints as well as result in unacceptable environmental damage.

China has the opportunity to forge a sustainable energy path without compromising its high rate of economic growth. Industrializing in an era of greater concern for the environment brings with it the opportunity to benefit from many advanced technologies for saving energy or reducing pollution impacts. These technologies are available or under development but not deployed in industrial countries because of limited markets, stranded costs, or strong vested interests. The size of the potential market in China allows faster deployment of cutting-edge technologies and would enable the country to become a leader in such deployment. The DRC and ERI's study shows that this objective is attainable without compromising economic growth, but doing so requires policies designed to rein in the growth rate of energy consumption to well below the growth rate of the economy. In particular, successfully attaining this objective calls for increased market orientation and incentives to bring about more aggressive deployment of these technologies for greater efficiency in the production, transformation, and end use of energy.

Without comprehensive policy actions now, China could lock its economy into a long-term, highly energy-intensive path. International data for the most industrial countries in the world show that, once major development and societal choices are made, those countries have been locked into particular narrow ranges of energy intensity for decades. One reason is that major energy-consuming equipment, such as power plants, industrial boilers, buildings, and so only have relatively long service lives (30 to 40 years). Another is that energy-intensive lifestyle patterns are very difficult to change once ingrained. As an example, planning for the development of compact cities, highways, and public transport to efficiently meet the mobility needs of urban dwellers has a greater chance of success if done before a society develops heavy reliance on private cars.

The timing of major decisions affecting investment in the key energy-consuming sectors is critical. The country urgently requires innovative policies, stronger enforcement procedures, and

> a streamlined implementation framework. The window of opportunity for shifting to a sustainable path is closing quickly because of rapid economic growth and the accumulation of low-efficiency capital stocks, which will still be in operation well beyond 2020, and because of increasing vested interests.

Sustained Consumption Growth, 1980–2000

During the two decades between 1980 and 2000, China experienced sustained growth in its consumption of energy.

Primary Energy Consumption More Than Doubled

As this report was being finalized, the government of China announced revisions of its gross domestic product (GDP) figures for the years from 1993 to 2004 and of its energy consumption data for the years from 1999 to 2004. (Appendix A presents the original and revised GDP and energy consumption data and elasticities derived from them.) The GDP revisions stem from surveys indicating that the contribution of the tertiary sector has been underestimated since the mid 1990s. These revisions seem to address longstanding concerns about the quality of GDP data and the underestimation of energy consumption during the late 1990s and early 2000s. Most observers of China's energy sector doubted the previously reported stagnation of energy consumption during the late years of the 9th Five-Year Plan (1996–2000) (about 1,300 mtce, with an average elasticity that was slightly negative). Instead, they imputed the low numbers to underestimation or underreporting of coal consumption during those plan years.

The revised data in table 2.1 (see p. 14) show that between 1980 and 2000 China's primary energy consumption more than doubled, rising from about 603 mtce to 1,386 mtce.[1] At the same time, GDP more than quadrupled, resulting in a 0.43 elasticity of energy demand—that is energy demand growth divided by GDP growth (hereafter *energy/GDP elasticity*). The lower elasticity in the 1980s and 1990s, to a large extent, reflected a successful shift away from the highly wasteful energy path of China's economy before the economic reforms that began in the late 1970s and the structural change in the economy that increased the share of less energy-intensive sectors in GDP.

Table 2.1 Primary Energy Production and Consumption, 1980–2000

	Production				Consumption			
	1980		2000		1980		2000	
Energy sources	mtce	%	mtce	%	mtce	%	mtce	%
Coal	442.4	69.4	928.8	72.0	435.2	72.2	938.0	67.7
Oil	151.7	23.8	233.5	18.1	124.8	20.7	321.4	23.2
Natural gas	19.1	3.0	36.1	2.8	18.7	3.1	33.3	2.4
Primary power[a]	24.2	3.8	91.6	7.1	24.1	4.0	92.8	6.7
Total	637.4	100.0	1,290.0	100.0	602.8	100.0	1,385.5	100.0

Source: China National Bureau of Statistics 2006 (abstract).

a. *Primary power* refers to hydropower, nuclear power, and renewable energy power generation. The conversion of hydropower and nuclear power to coal equivalents uses the average efficiency of thermal power plants in the same year. China began generating nuclear power in 1992, and in 2002, nuclear power accounted for 0.7 percent of total primary energy consumption.

Figure 2.1 (see pp. 16–17) presents the impacts of recent GDP and energy consumption revisions on the elasticity of energy demand to GDP[2] and on energy intensity.[3] It shows that although the revisions slightly affect past analyses of energy/GDP elasticity and energy intensity, they do not change past conclusions about these trends during the 10th Five-Year Plan (2001–05)—namely, that energy consumption is accelerating, that energy/GDP elasticity is rising, and that it may continue to do so in the absence of corrective policy measures.

Final Energy Consumption Grew at a Slower Pace

The definition of *final energy consumption* in the Chinese statistical system differs from the internationally accepted definition in two ways:

- It includes energy consumed in transformation uses—mainly electricity generation and petroleum refining.
- Its breakdowns by economic sector and fuel type are infected by some deficient accounting practices.

The most obvious example of such practices is that fuel consumption of cars owned by individuals or companies is not accounted for in the transportation sector but is included in the residential or the industrial sectors. For this report, adjustments have been made to estimate final energy consumption in line with accepted international definitions. The method and results of these adjustments are presented in appendix C.

Table 2.2 (see p. 18) presents the final energy consumption between 1980 and 2000 according to both the Chinese and the international definitions. In all cases, electricity is converted to million tons of coal equivalent on the basis of calorific value, not production equivalence value. From 1980 to 2000, final energy consumption grew 3.1 percent on average (about twice the world average) by the international definition, but only 2.7 percent on average by the Chinese definition. The difference seems to be consistent with the improved efficiencies achieved in transformation uses during the period.

The table shows also that the relative shares of the industrial, agricultural, and the residential and commercial (which are combined in Chinese definitions) sectors declined, although industry remained the largest consumer. The transportation sector more than doubled its share from 8 percent to 16.3 percent (after adjustment) and emerged as one of the key drivers of China's final energy consumption growth during the period. This trend was not well captured by the Chinese statistical system.

The structure of China's final energy consumption for 2000 shows a dramatic transformation over two decades: The share of coal, after adjustment, fell from 69 to 46 percent, while the shares of oil and electricity, respectively, increased from 18 to 26 percent and from about 6 percent to more than 15 percent. This transformation reflects patterns in the world's major industrial countries. With rising personal incomes and a more sophisticated economy, the relative shares of oil and electricity tend to increase: people can afford cars, air conditioners, and domestic appliances; commercial air travel takes off; factories use more electrically powered equipment; and a greater proportion of freight is moved by road. In line with these changes, the relative shares of the transportation and residential and commercial (buildings and appliances) sectors tend to increase. The shares of the residential and commercial and, more importantly, the transportation sectors in China are still low compared with more industrial countries. In 2000, they accounted respectively for about 20 percent and 16 percent of total energy consumption; compare these figures with 27 and 25 percent, respectively, in Japan. The shares of oil, gas, and electricity will therefore continue to rise if China follows the urbanization and transportation models of the more industrial countries.

Figure 2.1 Energy Intensity and Energy/GDP Elasticity, 1980–2005

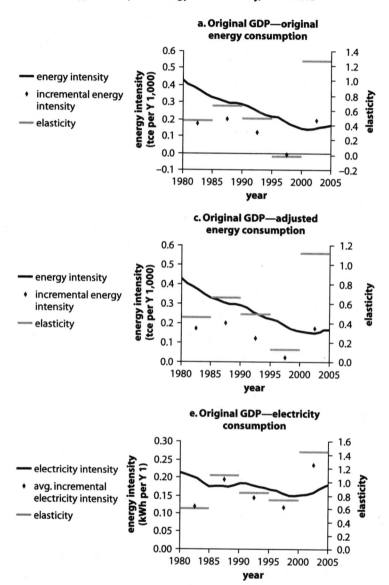

Source: Calculated and presented by World Bank team from China National Bureau of Statistics data.
Note: kWh = kilowatt-hour; tce = tons of coal equivalent.

Figure 2.1 Energy Intensity and Energy/GDP Elasticity, 1980–2005, Continued

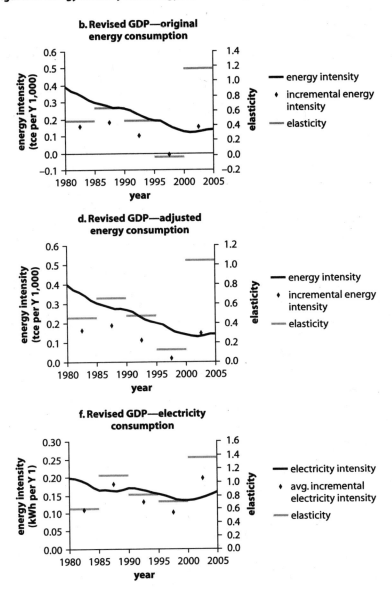

Table 2.2 Final Energy Consumption by Economic Sector and Fuel, 1980–2000

Economic sector	Chinese definition		International definition		Japan (2000)	Fuel type	Chinese definition		International definition		Japan (2000)
	mtce	%	mtce	%	%		mtce	%	mtce	%	%
1980											
Agriculture	34.7	6.0	31.1	6.9	48.3	Coal	335.6	58.3	311.6	69.0	11.0
Industry	392.1	68.3	277.0	61.2		Oil	89.5	15.6	82.1	18.2	59.0
Transportation	29.0	5.0	36.0	8.0	24.6	Gas	23.3	4.1	19.1	4.2	6.7
Residential	119.3	20.7	107.3	23.9	14.5	Electricity	116.1	20.2	28.3	6.3	22.2
Commercial					12.6	Others	10.6	1.8	10.5	2.3	1.1
Total	575.1	100.0	451.4	100.0	100.0	Total	575.1	100.0	451.6	100.0	100.0
2000											
Agriculture	42.9	4.4	40.2	4.9	48.3	Coal	421.6	43.4	382.1	46.2	11.0
Industry	664.8	68.5	489.2	59.2		Oil	267.9	27.6	213.2	25.8	59.0
Transportation	89.8	9.2	134.8	16.3	24.6	Gas	78.3	8.1	66.7	8.1	6.7
Residential	173.7	17.9	162.5	19.6	14.5	Electricity	154.1	15.9	128.3	15.5	22.2
Commercial					12.6	Others	49.2	5.0	36.4	4.4	1.1
Total	971.1	100.0	826.7	100.0	100.0	Total	971.1	100.0	826.7	100.0	100.0

Source: DRC and ERI, based on data from the China National Bureau of Statistics.

The Energy Statistics System Needs Rethinking and Overhauling
This quick review of the evolution of energy consumption at the primary and final user levels points to a seriously weakened system for tracking energy statistics (reasons are detailed in appendix C). The need to rethink and overhaul the system to improve its reliability and effectiveness is urgent, because major planning and policy decisions are based on it. There is a need for clear definitions and categorization of energy consumption to avoid arbitrary decisions and interpretation during data collection and reporting. Many of the Chinese experts who attended the dissemination workshop for this report, held in Beijing in June 2006, doubted the reliability of the system, stressed its deficiencies, and drew attention to the troubling appearance of some false statistical data, especially on coal consumption in the late 1990s and early 2000s.

This overhaul is also needed to align the system with internationally accepted standards, given China's increased integration with global energy markets. The issuance of the Implementation Decree for the Statistics Law on December 16, 2005, was a major step in addressing these problems. However, more remains to be done, as indicated in the succinct review provided in appendix C.

High Growth Forecast for 2000 to 2020

In the early 2000s, several major national and international institutions prepared energy projections for China for the period from 2000 to 2020. These projections assumed that the lower rate of energy consumption relative to that of GDP during the preceding 20-year period would continue.

This report uses the DRC and ERI's projections in their assessment of China's energy future because those projections were designed explicitly to help the government of China explore the issues, options, and uncertainties in shaping a sustainable development path. The DRC and ERI consulted extensively with national and international experts in energy technology and policy during the preparation of the projections and had access to detailed local information. Furthermore, the projections cover the range of energy consumption forecast in the studies by the International Energy Agency (IEA) and the U.S. Department of Energy (U.S. DOE), as well as other sources.[4]

The focus of the DRC and ERI's study was to explore policy options that China could implement to meet the same economic objective of quadrupling per capita GDP (from US$940 to US$3,760 in constant 2000 dollars) by 2020 with much less energy input per unit of output and greatly reduced pollution from energy production and use. These options would require

Table 2.3 Key Policy Elements Affecting the Projections of the DRC and ERI's Scenarios for 2020

Elements of a sustainable energy policy	Ordinary efforts scenario	Promoting sustainability scenario	Green growth scenario
Improvement in energy efficiency	Minimal priority	Moderate priority	High priority with a comprehensive system of financial incentives in place
Better environmental controls	Some use of desulfurization technology in power plants but not in the majority of plants by 2020	More stringent vehicle emissions standards in big cities; all power plants to have desulfurization equipment by 2020	More stringent environmental regulations and enforcement
Promotion of renewable energy and other clean energy sources	No explicit policy for development	Moderate promotion	Extensive promotion of renewable energy and gas substitution for fossil fuels
Key new technologies	None	None	Some application of integrated coal gasification combined cycle technology for electricity generation

Source: DRC 2000.

more stringent energy-efficiency and environmental policies and the deployment of new technologies. Table 2.3 summarizes the key differences in policies under the ordinary efforts scenario (business as usual) compared with those under scenarios with greater concern for energy sustainability.

Table 2.4 summarizes the energy projections of the DRC in collaboration with the ERI, along with those of the IEA and the U.S. DOE.[5] The DRC and ERI projections show that China's energy consumption would likely grow at 3.3 to 4.8 percent between 2000 and 2020, depending on the level of effort devoted to further improvements in energy efficiency and related choices of technology and modes of living. This growth would result in China's primary energy consumption in 2020 ranging between 2,430 and 3,300 mtce, or from 1.8 to 2.4 times the consumption level of 2000. Applying these growth rates to the revised energy consumption for 2000 leads to energy consumption of between 1,992 and 2,300 mtce in 2010 and of between 2,561 and 3,676 mtce in 2020.

Table 2.4 Projections of China's Primary Energy Consumption, 2000–20

Consumption and growth rates	DRC and ERI study scenarios			U.S. DOE study scenarios			IEA studies	
	Ordinary efforts	Promoting sustainability	Green growth	High	Reference	Low	2002	2004
Consumption level (mtce)								
2010	2,150	2,080	1,870	2,090	1,970	1,840	1,950	2,480
2020	3,300	2,900	2,470	3,220	2,790	2,430	2,560	3,170
Average annual growth (%)								
2000–10[a]	5.2 (4.5)	4.8 (4.1)	3.7 (3.0)	4.9 (4.2)	4.2 (3.6)	3.5 (2.9)	4.1 (3.5)	6.4 (6.0)
2010–20	4.4	3.4	2.8	4.4	3.5	2.8	2.8	4.6
2000–20[a]	4.8 (4.4)	4.1 (3.8)	3.3 (2.9)	4.6 (4.3)	3.9 (3.6)	3.2 (2.8)	3.5 (3.1)	4.6 (4.2)

Source: Data from DRC, ERI, and U.S. DOE.

Note: Figures are rounded for simplicity in presentation.

a. The growth rates in parentheses are based on the revised energy consumption of 1,385.5 mtce in 2000.

All the DRC and ERI's scenarios showed that the relative shares of the transportation sector and the residential and commercial sectors in the final energy consumption would increase significantly (see table 2.5, noting that the final energy consumption in the DRC and ERI's study is estimated using the Chinese definition). In particular, consumption by the transportation sector would increase substantially, from 9 percent of final energy consumption in 2000 to about 20 percent in 2020 (this share is higher if the international definition is applied). This increasing share of the transportation sector is a feature of the economic development and urbanization path that most industrial countries have followed. For example, in 2000, the share of energy used in the transportation sector amounted to about 16 percent of China's final energy consumption, after revision to meet the international definition. This share is substantially lower than the shares observed in industrial countries: 39 percent in North American members of the Organisation for Economic Co-operation and Development (OECD), 37 percent for European members, and 28 percent for Pacific members of the OECD.[6]

China still has substantial room for growth in this sector. Furthermore, all scenarios showed that freight would continue to dominate the transportation sector. Variations among the three scenarios in the projections of the sector's energy consumption were attributable primarily to assumptions about efficiency improvements in freight transport. These improvements are expected to result not only from improved technologies for trucks but also from a modal shift, placing relatively more freight on railways and waterways than on roads. Finally, all three scenarios showed rapidly rising ownership of private automobiles. Such high levels of energy consumption raise a major concern about energy sustainability.

Table 2.4 also shows that by 2020, with the appropriate policies and incentives (that is, the green growth scenario of the DRC and ERI's study), China could reduce its primary energy consumption by 25 percent relative to the ordinary efforts scenario. The promoting sustainability scenario, which represents a more moderate commitment of the government to improving energy sustainability, would result in a decrease in energy consumption by only 12 percent over the ordinary efforts scenario.

Recent Signs of an Unsustainable Energy Growth Path

Energy consumption during the 10th Five-Year Plan is higher than envisaged in all forecasts. Considering the revised energy consumption figures, the average elasticity of energy to GDP during the 10th Five-Year Plan

Table 2.5 Final Energy Consumption by Sector and Fuel, 2000–20

Sector	2000		2020[a]		Fuel	2000		2020[a]	
	mtce	%	mtce	%		mtce	%	mtce	%
Agriculture	42.9	4.4	59–80	3.1–3.2	Coal	382.1	46.2	559–836	29.3–33.7
Industry	664.8	68.5	998–1,345	53.4–54.2	Oil	213.2	25.8	605–827	31.7–33.3
Transportation	89.8	9.2	385–503	20.2–20.2	Gas	78.3	8.1	159–157	6.3–8.4
Residential and commercial	173.7	17.9	463–555	24.4–23.2	Power	154.1	15.9	434–487	22.8–19.6
					Other	42.9	5.0	148–177	7.1–7.8
Total	971.1	100.0	1,905–2,483	100.0	Total	870.6	102.0	1,905–2,484	100.0

Source: DRC 2000.

Note: The final energy consumption in the DRC and ERI's study is presented only according to the Chinese definition.

a. Low end of range represents the amount or share under the green growth scenario. High end of range represents the amount under the ordinary efforts scenario.

was slightly higher than one, more than twice the average elasticity during the three preceding five-year plans. At the same time, the energy intensity of China's economy—after a long period of steady decline—shows signs of stagnation and even slight increase (see figure 2.1). To reflect these developments, the IEA in 2004 revised its projections for China's energy consumption growth through 2020, from 4.1 to 6.4 percent. However, even this upward revision seems, in light of the 10th Five-Year Plan trend, to still underestimate the potential consumption growth (see tables 2.4 and 2.5).

Table 2.6 shows that following a rapid economic recovery from Asia's financial crisis during the late 1990s, China's primary energy consumption rose from about 1,386 mtce in 2000 to about 2,225 mtce in 2005, a growth rate averaging nearly 10 percent. This rate is more than two times that of the high economic growth period between 1980 and 2000. Energy growth accelerated during the 10th Five-Year Plan from about 3 percent in 2001, to some 6 percent in 2002, to about 15 percent in 2003, to 16 percent in 2004. That peak was followed by slightly reduced but still strong growth of 9 percent in 2005. Between 2001 and 2004, for the first time in two decades, the rate of primary energy growth outpaced the rate of economic growth. Electricity consumption statistics, which are

Table 2.6 China's Primary Energy Production and Consumption, 2000–05

Indicator	2000	2001	2002	2003	2004	2005
Total primary energy production (mtce)	1,290.0	1,374.5	1,438.1	1,638.4	1,873.4	2,063.0
Shares of production by source (%)						
Coal		72.0	71.9	72.3	75.1	76.0	76.3
Crude oil		18.1	17.0	16.6	14.8	13.4	12.6
Natural gas		2.8	2.9	3.0	2.8	2.9	3.2
Hydropower and nuclear		7.1	8.2	8.1	7.3	7.7	7.9
Total primary energy consumption (mtce)	1,385.5	1,432.0	1,518.0	1,749.9	2,032.3	2,224.7
Shares of consumption by fuel (%)						
Coal		67.7	66.7	66.3	68.4	68.0	68.7
Oil		23.2	22.9	23.4	22.2	22.3	21.2
Gas		2.4	2.6	2.6	2.6	2.6	2.8
Hydropower and nuclear		6.7	7.8	7.7	6.8	7.1	7.3

Source: China National Bureau of Statistics 2006 (abstract).

more reliable than energy consumption statistics, confirm these trends (see figure 2.1).

The more rapid growth of energy consumption during the 10th Five-Year Plan has caused concern in China and has attracted attention around the world. Energy demand is growing at higher-than-expected rates, certainly well beyond the objectives of the plan. Power shortages have plagued the high-growth belt in China and have triggered huge investments in power generation: 51 gigawatts of capacity in 2004, 66 gigawatts in 2005, and more than 200 gigawatts under construction in 2006. Most of this additional generation capacity has and will continue to come from medium-sized, relatively inefficient, polluting conventional coal-fired plants. In addition, oil imports during the 10th Five-Year Plan were well beyond most, if not all, forecasts.

The energy and electricity elasticities increased suddenly from an average of about 0.4 and 0.8, respectively, over two decades to more than 1.0 and 1.4 during the 10th Five-Year Plan. This increase suggests that China is heading toward a more energy-intensive path than even the ordinary efforts scenario forecasts. It indicates the need to establish a strong policy foundation for slower and cleaner energy growth if China is to build a sustainable future. Furthermore, international experience suggests that the role of energy as a factor of production in China could increase in the future and could contribute to even greater energy growth than anticipated.

A study by the IEA (2004, 329–33, box 10.2) assessed the contribution of energy to the economic growth of several developing countries that experienced high GDP growth in the 1980s and 1990s. In all the countries included in the study except China, the combination of capital, labor, and energy made a greater contribution to economic growth than did productivity increases.[7] The overall conclusion was that energy plays the leading role in the factors of production for countries in an intermediate stage of development. The study noted that at this stage energy-intensive industries are often the key drivers of economic growth but that as an economy matures, these industries become proportionately less important and countries develop more energy-efficient technologies that contribute to productivity increases.

There is some concern that China may not remain the exception that the IEA study indicates. China currently has a much lower per capita income level than some other countries in the study (Brazil, the Republic of Korea, and Turkey). If it were to continue following the model of industrialization of most developing countries (and industrial countries at earlier stages), energy intensity would likely increase with the growth of per capita income. Furthermore, China's most energy-intensive industries (chemicals, ferrous metals, oil refining, and the like) will likely raise their output substantially in the future. Instead of waiting for the economy to mature, China could chart a new course incorporating a strong emphasis on energy efficiency and cutting-edge technology. It could thus try to avoid the adverse impacts—in terms of pollution and oil import dependency—of the energy-intensive path of industrial countries that most developing countries are now following.

China's increased dependence on imported oil since the early 1990s is already a security concern, and that dependence will worsen if recent energy growth trends persist (see table 2.7 and figure 2.2). During the first decade of high growth (the 1980s), China was able to maintain quasi autarky in energy supply. However, by 1993, the country had become a net oil importer, and by 1995, its dependence on oil imports had increased to about 5.3 percent. This level of dependence was no cause for alarm. However, just five years later, in 2000, the growing demand of the economy had propelled China's dependence on oil imports to nearly 33 percent. Moreover, net oil imports grew at 14.2 percent on an average annual basis in those five years, to reach 143.6 million tons in 2005. At this level of dependence on foreign oil supplies, the country's energy security became a major concern for decision makers.

All projections of energy consumption compared with energy supply estimates indicate that China inevitably will become even more depen-

Table 2.7 China's Oil Consumption and Trade, 1990–2005

Consumption or trade position	1990	1995	2000	2001	2002	2003	2004	2005
Consumption (million tons)	114.7	160.7	225.0	229.5	248.6	271.9	317.2	330.1
Imports (million tons)	6.9	31.5	94.6	88.5	99.6	129.6	170.6	168.5
Exports (million tons)	30.3	23.0	20.7	19.1	19.5	23.6	19.0	24.9
Net imports (million tons)	(23.4)	8.5	73.9	69.4	80.1	106.0	151.6	143.6
Dependency on imports (%)	—	5.3	32.8	30.2	32.2	39.0	47.8	43.5

Source: Consumption data from China National Bureau of Statistics 2006 (abstract); import and export data provided by the General Administration of Customs of China.

Note: — = not available.

Figure 2.2 Net Oil Imports, 1990–2005

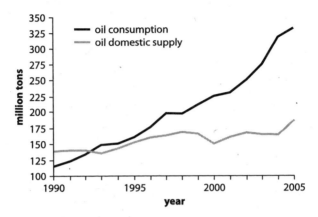

Source: DRC from government data.

dent on oil imports, even though its remaining exploitable oil reserves are significant. According to DRC estimates, these reserves amount to about 2.4 billion tons. Estimates of annual domestic oil output show production peaking at about 200 million tons in 2015 and thereafter ranging between 180 million and 200 million tons through 2020. In contrast, the DRC and ERI's projections indicate that by 2020 China's oil demand will be double or triple that amount, from 450 million to 610 million tons, in the low and high scenarios. As a result, the country's incremental oil demand between 2000 and 2020 would range from about 12 percent to about 28 percent of the incremental global oil demand projected by the IEA.[8]

Even if the current ambitious plans for synthetic fuel production[9] were to materialize, the contribution of these fuels to meeting oil demand would not likely amount to more than 20 million tons in 2020, less than 5 percent of the lowest estimate of oil consumption by that year. Therefore, China would increasingly rely on oil imports: the share of imports could rise dramatically, to between 50 and 60 percent of oil consumption by 2020. Furthermore, if recent trends continue, following the path taken by advanced market economies at earlier stages of industrialization (that is, if energy/GDP elasticity remains greater than 1.0), oil import requirements will exceed the highest current projections. The consequent pressure on world oil markets and the cost to China's economy would be enormous.

The major industrial countries also have experienced high dependency on oil imports, but some have managed to keep imports from increasing for several decades. For example, between 1973 and 2003, Japan maintained its oil consumption at about 250 million metric tons; during the same period, France actually reduced its oil consumption by about 23 percent, from about 120 million to 93 million metric tons. However, not all countries have been able to constrain imports. The United States, which has a more energy-intensive economy than either France or Japan, experienced an increase in both oil consumption and oil imports during the same period: oil consumption increased by about 8 percent, and the share of imports jumped from about 36 percent to nearly 64 percent because of declining production.

Despite the current high tension in oil markets, it is likely that global and regional markets would again adjust to meet the increasing demand, including China's high incremental imports. There is no consensus among experts about future equilibrium prices; however, it is likely that China would face tighter markets than industrial countries experienced in the past, even during the oil shocks of the early 1970s and early 1980s.

China's coal production could meet current and even higher consumption levels than projected for 2020, but those levels would not likely be environmentally sustainable. Energy projections show that coal would continue to account for more than 60 percent of primary energy consumption in 2020. Experts consider that the coal supply shortages and related reductions in electricity supply that occurred in recent years were a temporary phenomenon and not likely to present a long-term problem, particularly if transportation bottlenecks continue to be relieved. As box 2.1 indicates, massive reserves mean that China is likely to be able to meet coal demand through 2020 even if this demand turns out to be higher than envisaged.

Box 2.1

China's Coal Production Prospects through 2020

Under the DRC and ERI scenarios, even the scenario of high energy consumption would not result in an overall supply constraint for China. The country is currently the world's largest coal producer and second largest coal exporter. However, during recent years, especially from 2003 to 2005, coal supply seemed to lag the tremendous growth of demand. Power shortages during recent years resulted from the disruption of coal supplies, and Chinese producers failed in 2004 to honor contracts for export to power plants in the Republic of Korea (Martin-Amouroux 2005). But analyses by the DRC and ERI indicate that these disruptions are short-term bottlenecks and do not signal the decline of the coal industry in China.

China's coal reserves are abundant, according to most experts—despite the lack of agreement on their extent, owing to differences in definitions. The DRC and ERI estimate the country's exploitable coal reserves at more than 180 billion tons. On the one hand, those reserves could sustain the coal supply for about 65 years even if production were to rise to 2.8 billion tons—the maximum annual possible production, according to some Chinese experts. On the other hand, the development plans of the major players in the coal industry indicate that annual production could grow to more than 3 billion tons by 2020. However, some international experts contest these optimistic projections. More pessimistic projections indicate production deficits of 250 million tons in 2010 and 700 million tons in 2020 (Martin-Amouroux 2005). Beyond 2020, however, China could face constraints in coal supply if it were unable to improve efficiency in both the production and end use of coal as well as to promote fuel diversification more aggressively. In addition, some Chinese experts recognize other constraints on coal production, including the availability of water resources and the unsustainable environmental impacts of coal use.

The government of China is addressing the transport and coal-handling bottlenecks that have disrupted supply, especially to the fast-growing eastern coastal regions. Measures to remove these constraints have included, in the late 1990s and early 2000s, an increase in double-track railway length from 24 to 38 percent of the total length of the system and an increase in electrified lines from 13 to 29 percent of the total length of the system. In the coming years, investments by and on behalf of coal producers are likely to further improve the situation. Moreover, there are plans for an increase in the number of port installations for handling coal. These installations handled about 276 million tons of coal in 2003. Current expansion plans project an increase in handling capability to about 380 million tons in 2010 (Martin-Amouroux 2005).

China's relative level of dependence on coal is not unprecedented. Coal accounted for about half of global energy consumption in 1950. As recently as 1965, it still met about 40 percent of global needs, 53 percent of the former Soviet Union's, about 60 percent of the United Kingdom's, and some 64 percent of Germany's primary energy needs (BP Global 2006).

Source: Authors.

After 2020, however, China could face coal supply shortages, unless greater energy efficiency is achieved by effective, promptly initiated measures and unless there is further diversification of the fuel mix, particularly to include increased use of gas and renewable energy for power generation. However, the atmospheric concentrations of major pollutants from coal production and use have reached or are approaching unsustainable levels (see chapter 4). Unless major efforts are made to improve energy efficiency and deploy clean-coal technologies on a large scale, those concentrations will continue to rise roughly in step with growing consumption. Delay in addressing these fast-growing environmental problems could lead to irreversible damage to the atmospheric, agricultural, and forest environments. Urgency is the characteristic lacking in energy and environmental policy.

The present concentrations of levels of nitrogen oxides and sulfur dioxide have reached or even, in some cases, exceeded the caps that the government has set for these pollutants. In addition, the projections of pollution associated with the DRC and ERI's energy scenarios are all higher than the targets the government envisages for 2020. Even under the most optimistic scenario (green growth), sulfur dioxide emissions would be 100 percent higher—and nitrogen oxides emissions would be 50 percent higher—than the environmental caps under consideration (see chapter 4).

The Urgency for Policy Action

The need for policy action is indeed urgent. Current trends cannot be sustained, and the window of opportunity will not be open much longer.

Recent Trends in Energy Consumption Are Unsustainable

If China were to continue following the path of industrial and middle-income countries, its energy consumption would have to grow more rapidly than is currently envisaged by any reputable forecasters. Consumption would soon reach unsustainable levels, essentially because of the great size of the country's population and its economy. For example, if China were to attain the level of per capita energy consumption of Germany or Japan, its energy use would be equal to or slightly more than half of current world energy consumption. Without significantly greater reliance on cleaner technologies, such consumption volumes would have serious negative impacts on the environment, largely because of China's dependence on "dirty" coal. Such volumes would also put strong pressures on future regional and global oil (and probably gas) markets. Clearly then, present consumption trends are unsustainable.

In this connection, the higher energy intensity observed during the 10th Five-Year Plan is a major cause for concern. International data show that the economies of the more industrial countries tend to lock themselves into a narrow range of energy intensity for decades after making major development choices. For example, figure 2.3 shows that after the first energy crisis it took more than 30 years for the leading industrial countries to significantly reduce their energy intensity—mainly because large investments in equipment and processes have long life cycles and ingrained behavioral consumption patterns are often difficult to change. Furthermore, a large part of the reductions in energy intensity that were achieved probably stemmed from the deindustrialization of Western Europe and the United States, the related migration of some energy-intensive industries, and the continuing growth of the service sectors. However, China is in a position to take advantage of the most advanced technologies to leapfrog some stages in the growth of mature industrial countries, thereby achieving its full economic development with relatively less energy consumption and associated environmental impacts.

In summary, current trends are unsustainable, but China has a window of opportunity to break out of its highly energy-intensive growth path, to avoid the energy trajectories followed by the advanced industrial countries, and to adopt a path that is fundamentally more energy efficient and therefore environmentally sustainable. Such a path is in line

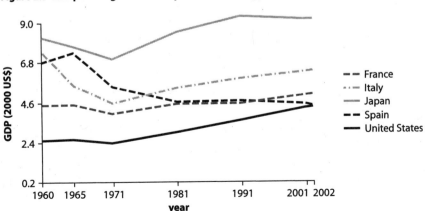

Figure 2.3 GDP per Kilogram of Oil Equivalent of Energy Use

Source: Study team estimates, based on the World Bank's World Development Indicators 2005 database.

with the government's policy objectives. However, shifting to this path demands urgent and coordinated action by China at the national and international levels. If China were to miss this opportunity, it could lock into an energy-intensive development path that could be costly for the country and for the rest of the world as well.

In China, the focus of concern about higher energy requirements is on their increasing cost to the economy: in 2005, energy costs in China amounted to 13 percent of GDP, compared with about 7 percent in the United States and slightly more than 3 percent in Japan.[10] This high cost could affect the competitiveness of China's economy in the long term. Chinese decision makers are also concerned about the security of supply, the volatility of global oil and gas markets, and the negative impacts of extensive use of coal on the local environment. Around the world, the focus is on the surge in China's oil imports and the resulting increased pressure on future oil supplies and prices, as well as the associated rise in greenhouse gas emissions and their impacts on climate change.

The Window of Opportunity Is Closing Rapidly

The successes of China's past policies in reducing power shortages and transforming the energy sector are well recognized. However, the recent trends are an early warning that the planning approach and current policies may have reached their limits and will not address the new challenges facing the sector. The planning approach is still fragmented, piecemeal, and strongly based on centralized command and control. In contrast, the energy sector is growing at a fast pace, driven by decentralized choices in an increasingly market-oriented economy. Sector operations are also growing in complexity because of greater integration with regional and global energy markets. Furthermore, there is a disconnect between the usual planning horizon (5 to 15 years) and the lengthy lead times required for structural changes owing to the long economic life of assets (30 to 40 years).

China faces a daunting development challenge. But there is an emerging consensus that the country's current unprecedented attention to energy and its increasingly damaging environmental impacts provide a unique opportunity to focus more strongly on sustainable development and innovative and workable solutions to meet the country's fast-growing energy needs. Such solutions are likely to require increased bilateral and multilateral cooperation to stabilize regional and international energy markets, taking full advantage of cutting-edge technologies to mitigate the environmental impacts of extensive coal use.

There is an urgent need for decisive actions to bring the energy sector back on a sustainable track. China has more than two decades of experience in piloting and assessing policy options. The time for action has arrived. China now stands at a crossroads. It faces major environmental challenges that require innovative thinking and strong political commitment to devise policies that will enable it to continue its unprecedented economic growth with less and cleaner energy. Meeting this challenge will also require more comprehensive international cooperation for technology transfer, if China is to meet environmental objectives in line with its national interests and those of the global community.[11]

Notes

1. This report focuses on commercial energy. However, biomass played and will continue to play an important role in China. See appendix B for a short note on biomass use in China prepared by Professor Wang Qenyi.
2. *Elasticity of energy demand* is the rate of energy growth divided by the rate of economic growth. It is a simple, easily measured, but very rough reflection of the relationship of energy to economic growth. It should be stressed that understanding the relationship between energy and economic growth requires more detailed and disaggregated analyses.
3. *Energy intensity* is the amount of energy input to produce a unit of economic output. At the macro level discussed in this report, it is expressed in terms of grams (tons) of coal equivalent of primary energy per yuan (per Y 1,000) of GDP (constant).
4. Projections published more recently are in the same range. For example, Li (2003) projects 2020 energy consumption of 2,800 mtce under a business-as-usual scenario.
5. The DRC and ERI projections and the U.S. DOE projections were prepared in 1999 and 2000. The IEA projections are more recent but do not have comparable scenarios; IEA's 2002 reference scenario projections of energy consumption were revised upward in 2004.
6. Unless otherwise noted, comparisons in the report are based on IEA (2002, 2004). Comparisons are useful, but they should be considered only as approximations because of the different energy accounting systems used by China (and other countries) and the IEA.
7. A study by the IEA (2004, box 10.2) concluded "In every country studied, except China, the combination of capital, labor, and

energy contributed more to economic growth than did productivity increases." However, questions were raised about China's productivity improvement, the largest in the group: "There are doubts about the accuracy of China's official GDP data. Many studies have concluded that the official statistics understate GDP and overstate growth rates" (IEA 2004, box 10.2).
8. In the reference scenario in IEA (2004), China's incremental consumption between 2000 and 2020 accounts for 18.5 percent of global incremental consumption.
9. Some Chinese experts commented that synthetic fuel production could be as high as 70 million tons by 2020. The report team considers these estimates highly overstated.
10. See assumptions considered in this calculation in appendix D.
11. This report focuses on China, but international cooperation should include also India, the other Asian giant.

References

BP Global. 2006. *Statistical Review of World Energy 2006*. London: BP Global. http://www.bp.com/productlanding.do?categoryId=91&contentId=7017990.

China National Bureau of Statistics. Various years. *China Statistical Yearbook*. Beijing: China Statistics Press.

DRC (Development Research Center). 2000. *Basic Conception of the National Energy Strategies*. Beijing: DRC

IEA (International Energy Agency). 2002. *World Energy Outlook, 2002*. Paris: IEA.

———. 2004. *World Energy Outlook, 2004*. Paris: IEA.

Li, Zhidong. 2003. "An Econometric Study of China's Economy, Energy, and the Environment to the Year 2030." *Energy Policy* 31: 1137–50.

Martin-Amouroux, Jean-Marie. 2005. "Le Charbon-Roi: Jusqu'a Quand?" *Revue de l'Énergie* 563 (January–February): 14–18.

CHAPTER 3

Reining in Future Energy Consumption

Key Messages

Maintaining energy consumption growth at a rate substantially less than economic growth will be more difficult in the future than it was in the last two decades of the 20th century, largely because doing so will require improved policies, greater reliance on markets, more rigorous planning, more efficient and targeted government interventions, and changes in the management of the sector. During the 1980s and 1990s, China's energy consumption grew at a slower rate than the economy because of centrally mandated reductions in energy waste combined with structural change in the economy. Now, however, decentralization, increasing market forces in the economy, diversified ownership of energy-using assets, and growing competitive pressures have substantially weakened the command-and-control approach that previously delivered recognized efficiency results. The projections of future energy consumption by sector show that energy use in transportation and residential and commercial will grow faster than in the other sectors owing to increasing freight movement (particularly by road), urbanization, and rising per capita incomes. This strongly decentralized consumption will make efficiency gains more difficult to achieve without proper economic and financial incentives, rigorous standards setting, and enforcement. It is therefore important for China to (a) evaluate best practices in the energy management of the world's market-based economies, (b) tailor these practices to China's unique characteristics, and (c) adapt the policy framework and the planning and management procedures to the more complex and fast-growing energy sector.

China's strong focus on energy efficiency in the preparation of the 11th Five-Year Plan (2006–10) is an important initial step toward reining in the growth of energy consumption, but this step requires clear definition of a baseline and detailed plans to achieve and monitor well-established targets. The 11th Five-Year Plan emphasizes the importance of establishing a resource-saving society. It includes a 20 percent reduction in energy intensity by 2010, which would amount to avoiding an increase of more than 600 million tons of coal equivalent (mtce) in China's energy consumption, more than the total energy use of any other country except Japan, the Russian Federation, and the United States. The 11th Five-Year Plan also notes the importance of establishing policy guidelines that stress energy savings. These guidelines should include several key elements:

- A clear baseline and a reliable energy reporting and accounting system to properly assess achievements
- A measurable outcome for each of the main energy-consuming sectors
- A focus on the largest, most inefficient consumers for maximum leverage in energy-efficiency gains over a relatively short period of time
- Reliance on a two-pronged approach with differentiated policies and measures targeting the efficient use of energy-consuming equipment and the acquisition of new capacity.

Energy-efficiency policies for five-year economic plans must be derived from a long-term energy-efficiency strategy designed to establish a new model of sustainable development, with the fundamental objective of keeping growth in energy consumption at a substantially lower rate than the growth of the economy. Energy consumption trends during the 10th Five-Year Plan (2001–05) indicate that China is embarking on a development path similar to the ones followed by industrial countries. Energy consumption during the plan years grew at a rate equivalent to or greater than the economy. Recent trends in energy consumption reflect the model followed by industrial countries during their early stage of economic growth. This model would be unsustainable for China in terms of use of resources, security of supply, and negative impacts on the environment. In its drive to build a resource-saving society, China needs to make energy efficiency the foundation of a long-term sustainable development model. This foundation should consist of

- The enforcement of efficiency regulations for existing energy-using capital stock
- The adaptation of advanced technologies for the most efficient energy use in all of China's energy-consuming sectors (leapfrogging), giving priority to those with the highest growth potential
- A strong communications program to involve the whole society in an endeavor that is vital to the energy security and well-being of future generations.

Achievements in Energy Efficiency

China's achievements in keeping growth in energy consumption at a sustainable level in the past are remarkable and have been the subject of considerable debate and study. From 1980 to 2000, China's primary energy consumption more than doubled, rising from about 603 to 1,386 mtce, at an average annual rate of 4.2 percent. At the same time, gross domestic product (GDP) quadrupled, growing at an average annual rate of 9.7 percent, indicating a 0.43 elasticity of energy demand to GDP growth for the 20-year period. This elasticity suggests a successful shift away from the highly wasteful energy path of the prereform era.

The energy intensity of China's economy has declined substantially since 1980, but the efficiency of major energy-consuming sectors is still well below international standards. The Development Research Center (DRC) has estimated total energy savings of 1,260 mtce between 1980 and 2000. These savings are nearly twice the incremental energy consumption over the 20-year period. They are implied savings compared with the energy consumption that would have occurred with an energy/GDP elasticity equal to 1.0, as experienced by Japan in its early stage of industrialization (an elasticity that was significantly lower than in other industrial countries). To explain these savings, analysts have advanced three main factors:

- A relative decrease in the weight of the energy-intensive sectors in the composition of GDP
- Increased energy productivity at the firm level
- Flawed statistics.

Numerous studies have attempted to evaluate the weight of each factor in the efficiency gains of the 1980s and 1990s.

A detailed analysis of studies on the relative weight of the factors affecting China's energy-efficiency improvements is outside the scope of this book. However, at least one study indicates that even if statistics in China are flawed to a certain degree, there is enough evidence of energy productivity increases and changes in the composition of GDP to overshadow the flawed statistics argument (Fisher-Vander and others 2004). During the period from 1980 to 2000, all economic sectors showed declines of 50 percent or more in energy intensity, with the overall energy intensity of the economy dropping from 4.29 to 1.46 tons of coal equivalent (tce) per Y 10,000 (table 3.1). Furthermore, as table 3.2 shows, the energy efficiency of the production of major energy-intensive industries increased in the range of about 20 to 60 percent, while the

Table 3.1 Change in China's Energy Intensity by End-Use Sector, 1980–2000
tce per Y 10,000, 2000 constant

Sector	1980	1985	1990	1995	2000
Agriculture	1.11	0.61	0.50	0.45	0.39
Industry	6.28	5.68	5.35	3.78	2.30
Construction	1.58	1.21	0.77	0.34	0.24
Transportation and communications	4.59	3.53	2.15	1.88	1.85
Commercial	0.78	0.33	0.47	0.40	0.39
Others	0.70	0.74	0.63	0.44	0.33
Average	4.29	3.28	2.89	2.18	1.46

Source: China National Bureau of Statistics various years.

Table 3.2 Improvements in the Efficiency of Key Energy-Using Equipment and Comparison with International Standards, 1980–2000

	1980			2000		
Item	Domestic average level	International advanced level	Gap (%)	Domestic average level	International advanced level	Gap (%)
Coal-fired power plant (gce/kWh)	448	338	+32.5	392	317	+23.7
Steel production (kgce/t)	1,201	705	+70.4	781	629	+24.2
Cement production (kgce/t)	203.8	135.7	+50.2	181.3	124.6	+45.5
Ethylene production (kgce/t)	2,013	1,100	+83	1,210	870	+39.1
Oil consumption of trucks (liter/10^2 ton-kilometer)	8.7	3.4	+155.9	7.6	3.4	+123.5

Source: ERI 2001.

Note: gce/kWh = grams of coal equivalent per kilowatt-hour; kgce/t = kilograms of coal equivalent per ton. Power supply was based on plants over 6 megawatts, steel data came from 75 large- and middle-sized enterprises, and cement data were based on large- and middle-sized rotary kilns.

truck transport sector registered a 14 percent increase in the efficiency of oil consumption. However, the gap between international levels of efficiency in the industrial countries and those of the major energy-consuming sectors in China is still substantial, ranging from about 24 to 124 percent.

The industrial sector remained the largest contributor to GDP, though its share was fairly stable over the 20-year period (table 3.3, see p. 40). Although the share of energy-intensive industries in total industrial output was still about 80 percent in 2000, the overall energy intensity of the industrial sector declined from 6.28 tce per Y 10,000 in 1980 to 2.30 tce per Y 10,000 in 2000. Part of the remarkable improvement is undoubtedly attributable to the growing share in the economy of a more efficient private sector and improved efficiency of large state-owned enterprises. At the same time, the relative shares of nonindustrial sectors in

Table 3.3 Sectoral Composition of China's GDP, 1980–2000

Sector	Share (%)				
	1980	1985	1990	1995	2000
Agriculture	29.9	28.2	26.9	19.8	14.8
Industry	43.9	38.3	36.7	41.1	40.3
Construction	4.3	4.6	4.6	6.1	5.6
Transportation and communication	4.5	4.5	6.2	5.6	7.4
Commercial	4.7	9.7	7.6	9.0	9.7
Others	12.7	14.7	18.0	18.4	22.2
Total	100.0	100.0	100.0	100.0	100.0

Source: China National Bureau of Statistics 2006 (abstract).

the composition of GDP increased—except for that of the agricultural sector, which declined from 30 to 15 percent.

Forecasts indicate that sustainability would require fundamental changes in policy. The energy scenarios explored the DRC and Energy Research Institute (ERI) study focused on providing analytical support in meeting the energy objectives of China's 10th Five-Year Plan and beyond. The forecast method focused on systematically thinking about the process of change in the energy sector to support sustainable economic growth. To rein in energy consumption, the reference scenario of the study (Promoting Sustainability) called for the implementation of the measures in the 1998 energy conservation law. The objective is to achieve by 2030—in all sectors—the energy-efficiency levels of the industrial economies.

The ordinary efforts scenario explored what could happen if the government did not achieve the measures of the energy conservation law and if the operating efficiency of major energy-using equipment did not reach the levels in the industrial economies. However, the focus of the DRC and ERI's work was on identifying a process of change that would aggressively pursue measures for expanding the economy on a sustainable basis (green growth scenario). This process includes the implementation of financial incentives and an energy pricing system to promote energy conservation; adoption and improvement of the measures of the energy conservation law; and as in the reference scenario, the attainment of advanced efficiency levels by 2030. The trends of the 10th Five-Year Plan clearly indicate the need for change if the government's sustainability objectives are to be achieved.

Concerns about Energy Consumption Trends of the 10th Five-Year Plan

The acceleration of China's energy consumption during the 10th Five-Year Plan, contrary to existing projections, reflects the development pattern of the industrial market economies.

Between 2000 and 2005, China's actual annual growth in primary energy consumption averaged 9.9 percent, and the energy/GDP elasticity averaged more than 1.0, which is more than double the elasticity of 0.5 that the existing energy projections implied for 2000 to 2020. As a result, by the end of 2005, actual energy consumption reached 2,225 mtce, equivalent to about 90 percent of the projected energy consumption of the green growth scenario for 2020 and 67 percent of the ordinary efforts scenario. From the experience of the 10th Five-Year Plan, China seems to be embarking on the same energy-intensive path experienced by the industrial market economies at the early stages of their industrialization.

The plateau in energy intensity of GDP corresponds to a decline in the share of investments for energy-efficiency improvements in total energy investment. Energy intensity has leveled off after a long period of decline, though it shows a slight increase in the early years of the new century. The easing of energy supply constraints is one possible reason for the decline in investments devoted to higher energy efficiency. This finding indicates the need for more aggressive policies and incentives for efficient energy use as supply constraints ease. If the energy consumption trends of the 10th Five-Year Plan were to continue for the next 15 years, energy consumption in 2020 likely would be double even the current high projections. Without prompt application of comprehensive energy policies to rein in energy consumption, China could be embarking on an energy-intensive path that could erode the security of its energy supply and have unsustainable impacts on the environment.

Bold New Directions of the 11th Five-Year Plan

"It should be made a basic national policy to save on resources," stated the committee that drafted the 11th Five-Year Plan (Communist Party of China Central Committee on Drafting the 11th Five-Year Plan 2005).

The proposals for the 11th Five-Year Plan, which began in 2006, represent a major shift for China in the direction of developing a resource consciousness that permeates all aspects of the economy and society. Departing from the past top-down and overly command-and-control approaches, the new policy advocates a strong focus on rapidity of achievement and on people-centered and resource-saving approaches. This strong statement follows from the 10th Five-Year Plan years, which saw energy consumption begin to take off in an unsustainable direction and which were marked by regional shortages of electricity and declining investments in energy efficiency.

The proposals have made energy efficiency a central concern, with adequate focus on qualitative and strategic directives but without the detailed action plans and policies and incentives required. The major target during the 11th Five-Year Plan—to reduce energy intensity by 20 percent—is challenging. It will require establishing specific programs with clear priorities, well-targeted policies and incentives, and substantial financial resources. However, it must be noted that the participants in the dissemination workshop for this report, held in Beijing in June 2006, have a more pessimistic view, as indicated in appendix E.

In 2004, China issued a medium- and long-term energy conservation plan. This plan is a strong sign of the government's recognition of the importance of energy saving for sustainable economic development. It also reflects some of the main themes in the proposals for the 11th Five-Year Plan, which pronounced the need to "upgrade the quality and benefit of economic growth," noting that "without vigorous energy conservation, the sustainable, rapid, harmonious, and healthy development of the economy cannot be supported" (NDRC 2004, 1). However, the targets of the long-term plan do not seem to be synchronized with those of the proposals for the 11th Five-Year Plan. There are no explicit links with the energy-efficiency targets for various types of energy-consuming equipment, nor is there a set of policies with an implementation plan for achieving them.

Reducing energy consumption by more than 600 mtce, as called for by the Five-Year Plan, is a daunting challenge. It will require a clear baseline from which to plan the efficiency improvements, especially because recent changes in GDP and energy consumption statistics cast strong doubt on the reliability of the reporting system and therefore on future achievements. For each of the major energy-using sectors, there needs to be a measurable efficiency outcome linked to specific targets required to achieve the outcome and the technologies necessary for efficiency improvements. It is also important to prioritize the efficiency targets, focusing on the most inefficient energy industries (mostly state-owned enterprises) for maximum leverage in achieving energy-efficiency targets.

Moreover, efficiency improvements should have two tracks—one for existing technology likely to remain in service in the medium to long term and another for the acquisition of new energy-consuming equipment that focuses on the plan's call for the most advanced technologies to meet development objectives. Finally, there is a need to place the objectives for the 11th Five-Year Plan in the overall context of a long-term vision for energy efficiency for the development of the energy sector.

The Closing Window of Opportunity for Reducing Long-Term Energy Intensity

The proposals for the 11th Five-Year Plan clearly recognize the closing window of opportunity that China faces for making fundamental changes to its development path. They suggest a strong sense of urgency for seizing the opportunity to develop in the first two decades of the 21st century. The window exists because millions of investors and consumers are all the time making (decentralized) decisions on capital stocks that will determine the future energy intensity of the economy. Those decisions will create the bulk of the energy-transforming and energy-using equipment that will exist in the national inventory of 2020: power stations, oil refineries, chemical plants, smelters, manufacturing operations, buildings of all kinds, domestic appliances, cars, trucks, trains, planes, and ships. One could add cities and highways to that list, because they indirectly have a strong influence on energy intensity.

The opportunity exists to chart a course along a new low energy-intensive path, rather than following the route of the major industrial market economies. Time is of the essence, because charting a new course can be achieved only by quickly and strongly modifying those investment and consumption decisions. The window of opportunity is closing rapidly because energy-using capital stocks, ranging from giant power stations to home air conditioners, are growing very quickly and because the public, corporate, and private investment decisions about them and about their use are currently being made without the reliable information, proper economic signals, and financial incentives that are needed to steer the economy and society along an energy-saving path. The longer the economy and society go without these data, signals, and incentives, the more frequently investment choices will be made in favor of relatively wasteful energy-transforming and energy-using equipment based on current technology.

Developing a resource-conscious society will not come easily; it requires strong and continued advocacy supporting a comprehensive

Figure 3.1 China's Per Capita Energy Consumption and GDP Compared with Selected Countries and the World Average

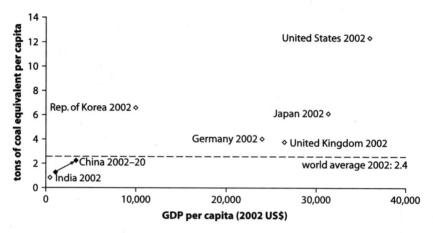

Source: World Bank study team.

and coordinated development vision, accompanied by a comprehensive energy policy. Taking account of the key elements for meeting the objectives of a resource-saving energy sector, the following paragraphs discuss an alternative energy development model that is in line with the objectives of the 11th Five-Year Plan; explore the major drivers of energy consumption through 2020; and provide examples, by end-use sector, of some of the priorities for improving energy efficiency. The discussion draws repeated attention to the symptoms of the current problem: high and possibly rising energy elasticity and an energy intensity that may have stopped falling and is high relative to the industrial countries.

China's per capita income and energy consumption indicate that the country is at a relatively early stage of energy development, with enormous potential for growth in the next decades. As figure 3.1 indicates, the country's per capita energy consumption, at 1.71 tce, is still lower than the world average of 2.44 tce (2005 figures) and would remain so in the energy projections of the DRC and ERI, which suggest a per capita consumption of about 2.2 tce in 2020. However, because of the country's large population, even growth at a demand elasticity less than was experienced by any of the industrial market economies (that is, less than 1.0) would make China's total energy demand by 2020 greater than that of the United States and of all of the European countries of the Organisation for Economic Co-operation and Development (OECD) after their

Figure 3.2 China's Projected Growth Path of Energy Demand, 1980–2020, Compared with That of Other Countries

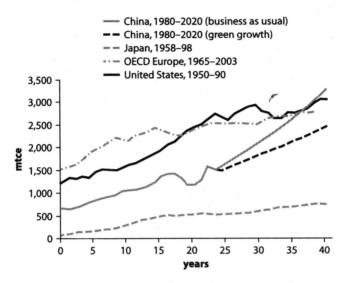

Sources: Data from BP Global 2004, DRC, Government of Japan, and U.S. Department of Energy.

40-year period of high growth (figure 3.2). For example, the energy consumption of the United States grew from 1,300 to 3,300 mtce as real GDP quadrupled between 1950 and 1995. The projections indicate that China is likely to achieve such an increase in less than half that time.

With ordinary efforts (business as usual), accelerating industrialization may require more energy than projections to date suggest. The industrial market economies have followed a development model that begins with an initial steep rise in energy intensity, with heavy industry dominating energy consumption. The duration of this phase varies by country, but in most cases it has lasted at least 20 years. After the country reaches a peak intensity level, a period of declining energy intensity sets in, which can last several decades.[1] Then, following a period of decline, countries tend to lock into a narrow range of energy intensity and remain there for several decades because of limits to energy-efficiency improvements in large capital stocks, which can have an economic life of several decades. Even Japan, recognized as the world's most energy-efficient large economy, had an energy/GDP elasticity of 1.14 during its rapid industrialization in the 1960s. The low energy/GDP elasticity for China recorded during the 1980s and 1990s stemmed from the specific characteristics of the country's past development patterns.

Figure 3.3 China's Energy Intensity and Major Development Periods, 1954–2005

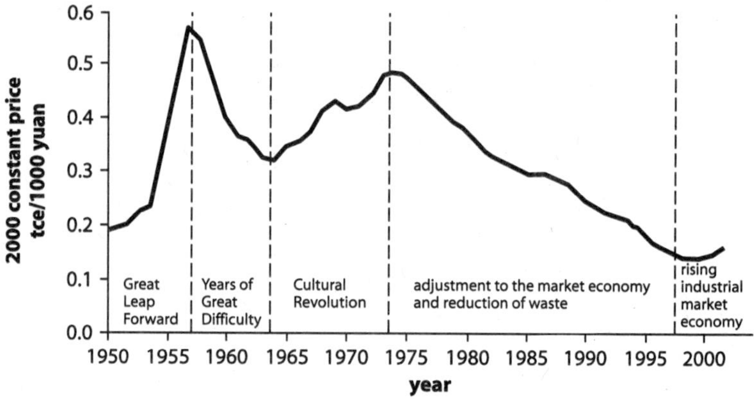

[Chart: Y-axis: 2000 constant price tce/1000 yuan, ranging 0.0 to 0.6. X-axis: year, 1950 to 2000. Labeled periods: Great Leap Forward; Years of Great Difficulty; Cultural Revolution; adjustment to the market economy and reduction of waste; rising industrial market economy.]

Source: Authors.

China's energy intensity rose and fell in response to different economic development policies that the country experienced during the past four decades (figure 3.3).[2] The long, steady, decline in the energy/GDP elasticity between 1980 and 2000 reflects a reduction of formerly high levels of energy waste. That waste resulted from growth models based on self-reliance and the development of small-scale industries in the countryside, villages, and small cities. Most of the time, these industries were highly inefficient and polluting. The energy savings of the first two decades following the open door policy stemmed from the restructuring of the highly fragmented industrial sector and gradual energy price increases as the market orientation of the Chinese economy increased.

The sharp contrast between actual consumption and the projected trend during the past five years—with an energy elasticity close to 1.0— raises a justifiable concern that China could be embarking on the same energy path that industrial market economies have followed. As table 3.4 suggests, energy elasticity could rise even further, entailing significantly higher energy consumption than is currently forecast. If China's energy consumption were to increase at the same rate as economic growth, energy consumption in 2020 would amount to more than 5,000 mtce, with consequent threats serious to the environment and security of supply.

China could reduce energy intensity by adopting a new development path. It does not necessarily have to follow a steeply rising curve of energy intensity to reach higher levels of per capita income, as the

Table 3.4 Energy Demand/GDP Elasticities of Major Industrial Countries, 1961–2002

Country	1961–72	1961–2002
France	1.24	0.93
Italy	1.87	1.14
Japan	1.14	0.96
Spain	1.24	1.27
United States	1.06	0.59

Source: Study team estimates based on country statistics.

industrial market economies have done (figure 3.4). Instead, the country could develop along a new path that would improve the economic well-being of its people while avoiding the adverse impact of rising energy intensity on the economy and the environment (Berrah 1983). However, the window of opportunity for change is closing because of the high growth of consumption, the rapid accumulation of outdated (and even obsolete) energy-transforming and energy-consuming equipment, and the absence of detailed policies and targeted incentives to steer the country's development toward a less energy-intensive path. Incorporating energy efficiency in decisions on lifestyle choices (the "energy-conscious society")—such as the choice of mass public transportation systems over widespread dependence on the personal automobile—and incorporating the most advanced energy-efficiency technologies in all new investments (technological leapfrogging) are necessary to put China's energy sector on a sustainable path.

Figure 3.4 Tunneling a Less Intensive Energy Path to Higher Per Capita Income

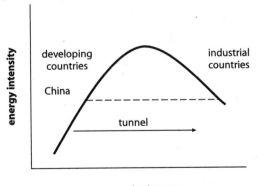

Source: Authors.

The alignment of China's economic development along a less energy-intensive path than any other country has experienced in an early phase of industrialization will require long-term vision, innovative approaches, and strong policies to drive the economy and the society toward resource and energy savings. For existing assets, the emphasis should be on well-established measures to further reduce energy consumption, such as the closure or technical retrofitting of inefficient industrial plants and better enforcement of existing energy-efficiency policies and standards. The ability of new energy-consuming assets to reduce energy intensity can be that much greater by recourse to technological leapfrogging and the institution of major lifestyle changes. The needed reductions in energy intensity will not be easy to achieve, but they are essential to sustainable development.

The key drivers of China's future energy consumption are growing urbanization, rising per capita incomes, and the continued development of China as a major global manufacturing base. The following paragraphs examine the scope to rein in energy consumption in four principal subsectors: power generation, industry, transportation, and residential and commercial building.

Projections show that the consumption of energy in the transportation and building sectors would increase much faster than in other sectors between 2000 and 2020. Continuing urbanization (entailing new housing, more commercial and public buildings, and greater demands for mobility) and rising personal incomes are driving up energy consumption. People with rising incomes seek better housing and greater indoor comfort (heating and cooling). They want to have their own cars and to travel at will. The projections for energy use for transportation and buildings are based on China's degree of urbanization increasing from about 26 percent in the base year (2000) to between 53 and 58 percent by 2020, depending on the scenario. These projections represent a major shift because, on average, China's urban residents consume about 3.5 times as much energy as rural dwellers.

Despite a structural shift toward a greater share of the service sector and of less energy-intensive industries in the economy, projections indicate that heavy industry would likely still account for about 80 percent of the projected energy consumption of the industrial sector in 2020. This finding is attributable in large part to globalization and China's continued role as the factory for the world, owing to its large, low-cost labor market and major scope for productivity improvements. This role essentially transfers the demand of energy-intensive sectors from other countries to China. Both urbanization and the growth of the manufacturing sector are likely to increase the demand for energy in the transportation sector. Greater production for the global market will entail an expanded transportation infrastructure and mobility of goods.

Power Generation

The power subsector still has significant scope for efficiency improvement through lower specific consumption by demand-side management and further efficiencies on the supply side, particularly in thermal power generation.

A study by the World Bank's Energy Sector Management Assistance Program concluded that, with adequate policies and targeted incentives, demand-side management could reduce China's total 2020 electricity needs by about 220 terawatt-hours and subsequently the capacity needed to meet them by more than 100 gigawatts (Gu, Moskovitz, and Zhao 2005). Tapping even half of this potential would reduce coal consumption by about 37 mtce in 2020. The vast potential of demand-side management remains largely underdeveloped in China because there are very powerful barriers to customer and utility investment in energy efficiency.

Well-known barriers found in other countries that are also present in China are the lack of capital for investment in energy savings; inadequate information about efficiency potential; and electricity prices that do not reflect the full direct and indirect costs of power generation, transmission, and distribution. In addition, there are several other barriers unique to China. They include the lack of a legal basis for the effective adoption of demand-side management policies as well as of adequate, stable funding mechanisms. International experience shows that public or utility funding for demand-side management is critical to its success. Much of China's industrial base is in transition, coping with enormous economic change and restructuring. The resulting uncertainty makes it difficult for enterprises to commit funds necessary for energy-efficiency projects. At the same time, China's approach to pricing at both the generation and retail levels is inadequate, because there are no incentives for the grid companies to promote energy efficiency and sound load management.

China's current method of setting electricity tariffs does not give consumers correct signals to invest in demand-side management and discourages grid utilities from doing so. Some of the energy-efficient products that China produces for the domestic market, such as compact fluorescent light bulbs, are of low quality, even though China exports large quantities of high-quality bulbs. Furthermore, China has a shortage of specialized professionals trained in designing and marketing demand-side management products.

China's suboptimal dispatch of existing power generation capacity is another major inefficiency factor and reflects inadequate pricing. Most efficient plants do not operate at full capacity, especially during off-peak hours. Often power is produced from less efficient and more polluting power plants in order to meet contractual arrangements that are based on an energy price (on kilowatt-hours) and a minimum number of dispatch hours rather than on capacity price (on kilowatts) and fuel cost (on kilowatt-hours) arrangements. Suboptimal dispatch has resulted in an increase in coal consumption estimated at between 25 and 35 mtce per year during the past five years, or about 6 percent of the total coal required for power generation.

The performance of coal-fired power plants is critical because they account for about 70 percent of China's current power generation capacity and about 41 percent of China's primary energy consumption. This share of coal in power generation is likely to remain about the same through 2020. Therefore, the priorities of any efficiency program must be the application of the best available technologies for coal power generation and the maximum use of efficient, low-polluting plants. During 1980 to 2000, China improved the gross efficiency of coal use in power generation by more than 25 percent. However, most of this improvement did not come from the adoption of advanced technologies. Instead, it resulted from phasing out small, inefficient units (smaller than 50 megawatts) in favor of newer, larger, and more efficient conventional plants. Nevertheless, China's thermal power generation still requires about 24 percent more primary energy consumption per kilowatt-hours of output than the average for the leading industrial nations.

A more disturbing trend is the low level of improvement in the efficiency of electric power generation between 2000 and 2005. The increase was only 3 percent. This very small improvement is mainly attributable to a large increase, during the period, in the number of new, relatively small, inefficient power generation units installed. The purpose of these units was to alleviate power shortages that resulted from a significant underestimation of electricity demand growth under the 10th Five-Year Plan. However, the lack of substantial efficiency gains also reveals the increasing difficulty of achieving further improvements without technological leapfrogging.

For China to achieve, by 2020, gross energy efficiency comparable with that of France, Japan, or the United Kingdom in 2000, the energy consumption of all additional generating units in China's power system during the next 15 years should not exceed 290 grams of coal equivalent (gce) per kilowatt-hour. Hence, all capacity additions not yet under construction should be at least supercritical, ultrasupercritical, or integrated coal gasification combined cycle (IGCC) plants. However, current policies for the development of the power subsector are far from promoting deployment of these technologies on a significant scale. For example, during the first four years of the 10th Five-Year Plan, about 32 percent of the added capacity was in relatively small units of between 100 and 300 megawatts. The average consumption of these units was about 350 gce per kilowatt-hour. Furthermore, another 30 percent of the added capacity included units generating less than 100 megawatts, with average consumption of 390 gce per kilowatt-hour. This pattern is also reflected in the 200 gigawatts of capacity currently under construction.

There are two main reasons for the disturbing trend of continued investment in smaller, less efficient generating capacity. First, power shortages dictated the need for hasty capacity additions. Second, investors favored smaller units, despite their lower efficiency levels, because of their low initial cost and their short construction time. No consideration was given to possible long-term environmental impacts. If the focus remains primarily on investment costs with no internalization of environmental costs, the decision-making process will continue to favor investment in units that have a low initial cost without regard for their overall lifetime cost. Without major policy changes, the result could be an increasing proliferation of units with lower efficiency and higher levels of pollution than others available on the market. Investments in such technologies imply a strong preference for the present to the detriment of future generations and are therefore inimical to sustainable development.

A striking example of short-term vision in power planning is the lack of sufficient attention to investments in available advanced technology. For example, although the estimated capital cost of IGCC plants is nearly double that of the conventional, supercritical, and subcritical power plants that China produces, IGCC units require only about 272 gce per kilowatt-hour compared with 300 to 355 gce per kilowatt-hour for supercritical and subcritical units ranging in size from 300 to 600 megawatts. However, an investment analysis that considers externalities—that is, the social costs of coal use (environmental damages attributable to local pollution)—during the whole life cycle of the assets would find IGCC units economically justified with only a 15 percent reduction in capital costs (see appendix F, which examines life-cycle costs of power generation investments with conservative estimates of environmental costs factored in). The focus on only the high capital cost of IGCC units, rather than their overall lifetime costs, has delayed the construction of a 400-megawatt pilot unit that has been under consideration for more than 10 years.

A reduction in the capital cost of IGCC technology by 15 percent is well within the capability of Chinese industry. The country has a strong track record of adapting international technologies and producing them domestically at a lower cost. For example, the costs of conventional coal technologies per kilowatt installed in China are about 30 to 50 percent lower than those in other countries. If China were to add IGCC units to meet half of its new coal-fired power capacity requirements between 2010 and 2030, coal consumption for power generation would decline by around 11,000 mtce. Chapter 4 discusses options for financing the deployment of advanced clean technologies in China.

Industry

In the industrial sector, the greatest potential for efficiency improvement is in a few large, energy-intensive industries. The DRC and ERI's projections show that by 2020 industry would continue to account for about 60 percent of final energy consumption, or between 1,398 and 1,924 mtce. Energy-intensive industries would account for about 80 percent of the final energy consumption of this sector—that is, 48 percent of the country's total final energy consumption. All energy-intensive industries would nearly double their output between 2000 and 2010, with the highest growth in the ethylene and paper industries (figure 3.5). The DRC has estimated that the energy-savings potential during 2000 to 2020 would be about 250 mtce, equivalent to between 25 and 54 percent of the incremental consumption projected for the industrial sector during the period.

Although China has improved the efficiency of its energy use in industry, there is still a wide gap between the sector's efficiency and international efficiency standards. Closing this gap is important for the existing stocks, because China is already the largest manufacturer of many of the world's most energy-intensive products, including steel, cement, and ammonia. For example, China currently requires about 21 percent more energy to produce a ton of steel than international standards for the industry. The consumption of the country's cement and ethylene industries requires 44 percent and 70 percent, respectively, more energy than international standards for these industries.

Figure 3.5 Energy-Intensive Industries to Double Output by 2020

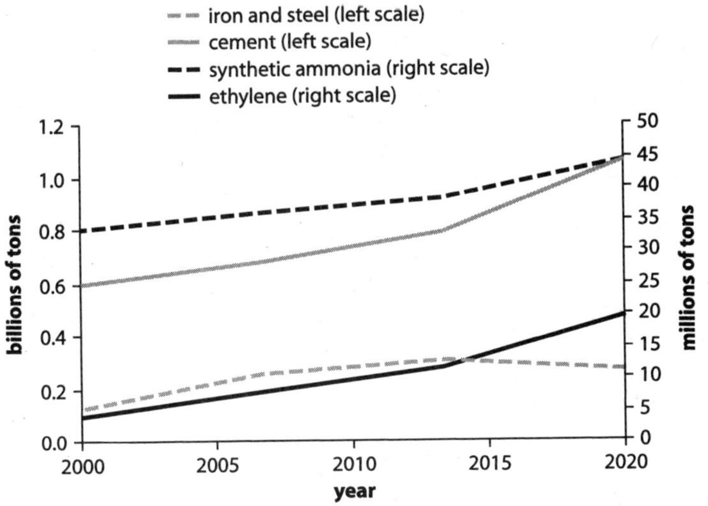
iron and steel (left scale)
cement (left scale)
synthetic ammonia (right scale)
ethylene (right scale)
billions of tons
millions of tons
1.2
1.0
0.8
0.6
0.4
0.2
0.0
50
45
40
35
30
25
20
15
10
5
0
2000
2005
2010
2015
2020
year

Source: Data provided by DRC.

Achieving greater energy efficiency in the industrial sector will require more aggressive and better-targeted efforts to address the specific conditions of each industry. These efforts need to combine higher, strictly enforced standards with proper incentives to tap existing potential, with the primary focus being on restructuring and renovation in low-growth industries and on technological leapfrogging and innovation in high-growth industries. Box 3.1 (see p. 54) provides a case study of a successful effort.

Transportation

Under the ordinary efforts scenario, the transportation sector would increase its energy consumption faster than any other sector, but there is substantial scope for energy savings through a combination of shifts in modes of transportation and in the use of highly efficient trucks and cars.

The energy projections show that the size of China's fleet of transportation vehicles would soar between 2000 and 2020 from 16 million to nearly 94 million vehicles.[3] Correspondingly, transportation fuel demand would more than triple, from 124 mtce to as high as 535 mtce, amounting to about 16 to 17 percent of primary energy consumption in 2020. Given the low level of existing stock in the transportation sector, there is enormous opportunity to shape lifestyle patterns to meet the mobility needs for goods and people through the most efficient modes and to leapfrog

> **Box 3.1**
>
> **The Potential Benefits of Technical Retrofitting: The Case of the Jinan Iron and Steel Group Corporation**
>
> Case studies of energy consumption in Chinese industries have shown that those with a combination of outdated equipment, inefficient energy use, and financial difficulties but a relatively modern management perspective can benefit from retrofitting of existing capital stock. An example is the Jinan Iron and Steel Group Corporation. Founded in 1958, this medium-sized steel enterprise produces about 3 million tons annually (2001 figure). The company has benefited enormously from several energy-efficiency measures—including dry coke quenching, off-gas utilization, and increases in the continuous casting ratio of up to 100 percent. As a result, the company's steel output increased by nine times between 1983 and 2000, while its energy consumption only slightly more than quadrupled. During the 8th Five-Year Plan (1991–95), the company invested US$101 million in energy savings, or 36 percent of its total investment over the plan period. According to a 2002 report to the State Economic and Trade Committee, the investments produced estimated profits equivalent to US$229 million.
>
> *Source:* Chen, Farinelli, and Johansson 2004.

to the most advanced and efficient cars and trucks to further reduce the incremental demand for oil, all of which is imported at the margin.

Because of the rapid development of the economy and the growth of its automotive stock, a gradual approach to key decisions in transportation sector development could undermine the sector's long-term sustainability. A greater focus on public transportation for passengers and railways and waterway infrastructure for freight, combined with leapfrogging to higher fuel economy standards for vehicles of all kinds, could make a dramatic difference in the sector's energy efficiency. Indeed, nearly 60 percent of the difference in primary energy consumption of the sector in 2020 between the ordinary efforts and green growth scenarios is due to a shift in modes of transportation, with relatively less reliance on individual cars, more freight transport by rail and waterway than on highways, and more energy-efficient trucks and cars.

Projections show that freight transport would continue to dominate energy consumption in China's transportation sector, accounting for about 60 percent of its projected consumption in 2020. Although waterways would then carry more than half of China's freight traffic, they would account for only about 22 percent of energy consumption in freight transport. In contrast, transport of freight by trucks would account for only 14 percent of transport in terms of distance but about 50 percent of energy consumed for freight transport.

The sooner the government makes the decision to support higher energy efficiency and the shift in modes of transportation, the better for China's energy security. For example, if China were to adopt the 2010 U.S. corporate average fuel economy standards for trucks without delay, rather than gradually, the savings in fuel consumption for trucks in 2020 would amount to about 44 percent of projected consumption, ranging from 36 percent for heavy-duty trucks to 58 percent for medium-duty trucks.[4] Currently, there is a huge gap between the energy efficiency of China's truck fleet relative to international standards (122 percent). Given this low efficiency and the importance of freight transport in total energy demand of the transportation sector, one high-priority policy is improvement in the energy efficiency of the truck fleet (see figure 3.6).

Although the passenger vehicle fleet is still very small, projections like those cited show that it could grow dramatically, depending on the choices the country makes today—mainly in the design of cities and the role of public transportation systems (figure 3.7, see p. 56). As China

Figure 3.6 Projected Truck Fleet by 2020

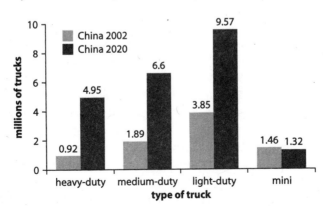

Source: Data provided by DRC.

Figure 3.7 Increase in Vehicle Population

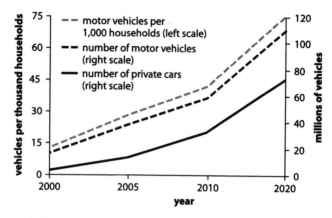

Source: Data provided by DRC.

industrializes further and as incomes rise, the country will face the choice of moving toward an automobile culture, as in the United States, or a mass transit culture, as in Japan and some European countries. Currently, in most Chinese cities, car ownership is only about 50 per 1,000 persons, compared with more than 700 per 1,000 persons in the United States. However, sales of passenger cars are rising rapidly and will continue to accelerate over the next 10 to 15 years, posing an enormous challenge to cities. As these sales increase, lifestyles could become more dependent on cars. If this dependence reaches a critical mass in terms of personal preferences and in terms of infrastructure of all kinds that supports automobile use, the pattern will become more difficult or even impossible to reverse.

A recent urban transportation strategy prepared by the World Bank indicates that despite the rapid growth, most urban households do not yet have cars, and the mobility needs of these households are not adequately met (World Bank 2006).[5] However, rapid motorization has left little time for governments to prepare strategically effective action plans. Few cities have the financial capability to develop mass transit systems, particularly underground systems, which require large investments and long lead times. Without policy action, an automobile-dependent, low-density suburban lifestyle has been taking shape in some major cities. An alternative lifestyle based on compact cities with mobility provided by mass transit systems using clean forms of energy and limited reliance on personal cars requires coordinated, long-term planning and investment. Many studies have been advising the same course of action, without great success at the national or local levels, except in a very few cities. This lack of success is attributable to a tendency for planners to base urban development on models and ways of life already adopted by the industrial countries.

Optimum strategies for sustained and sustainable growth are well beyond the scope of this report. Nevertheless, it is possible to draw some preliminary conclusions based on international experience. For example, the promotion of mass transit and public transportation systems does not drastically reduce car ownership. However, if combined with proper fuel taxation and mass transit infrastructure, as in Japan, it often reduces car usage and, consequently, oil product consumption. Furthermore, China's car manufacturing industry is at a relatively early development stage. The country, therefore, is well placed to add new stock that uses the most advanced technologies for fuel efficiency. Box 3.2 (see pp. 58–59) illustrates one way to achieve environmental sustainability without compromising the automotive development strategy (World Bank 2006, 3, 15).

Residential and Commercial Building
Residential and commercial buildings have large scope for improvements in energy efficiency, mainly through improvements in heating and cooling systems and the enforcement of building codes (box 3.3, see pp. 60–61). Past energy-savings measures focused on industry and overlooked the avoidable energy losses in buildings. As a result, only about 2 percent of urban building space (231 million square meters) has been constructed with the required energy-efficiency measures, such as thermal insulation of building envelopes (DRC 2004). Every year lost in enforcing existing standards and developing new ones locks in some 700 million to 800 million square meters of energy-inefficient building space that will be in use for decades. Space heating currently accounts for 54 percent of energy use in the buildings and comes mainly from coal-fired systems. The energy waste results from energy-inefficient designs and materials used for construction. In particular, central heating systems are still based on designs from the 1950s, which do not allow adjustment of heat levels by the consumer (World Bank 2001).

Ongoing studies by the World Bank and Chinese authorities indicate that residential buildings in China require at least twice the energy per square meter to maintain room temperature at the same level as buildings located in similar climates in Europe and the United States. Even most new buildings (about 95 percent) are not compliant with the government's existing energy-efficiency codes.

Box 3.2

Reconciling Automotive Industrial Policy and Energy Sustainability

Analyses presented in the World Bank (2006) report "China: Building Institutions for Sustainable Urban Transport" estimated the levels of fuel consumption and greenhouse gas emissions under three scenarios:

- The *road ahead scenario* assumed the current growth rate of motorization and current fuel prices, with conventional gasoline vehicles as the dominant vehicular technology.
- The *oil-saved scenario* assumed the current growth rate of motorization, but with hybrid electric vehicles gaining 15 percent of the market by 2010 and 50 percent by 2020. Apart from conventional gasoline vehicles, the market penetration of compressed natural gas and small electric vehicles would be relatively high. This scenario further assumed a higher share of smaller vehicles in the fleet and higher oil prices, which would reduce average vehicle use. The assumed advanced technologies for this scenario are those already adopted or being adopted by some industrial countries.
- The *city-saved scenario* went one step beyond the oil-saved scenario. It assumed a lower level of vehicle use resulting from the development of compact cities in addition to improved public transportation services. In this scenario, small and fuel-efficient vehicles would play a considerable role in reducing fuel consumption and emissions. Hybrids, together with small electric and compressed natural gas vehicles, would dominate the market, with conventional gasoline vehicles constituting only 30 percent of the total market. There also would be an increase in the use of small electric cars (for example, smart cars) by about 30 percent.

The simulation results for the various scenarios, in the figure below, show the enormous reductions in energy use and greenhouse gas emissions that are possible without compromising the development of the automotive industry. Although the estimates are subject to further refinement and debate, they would result in a huge potential payoff from the adoption of the city-saved scenario.

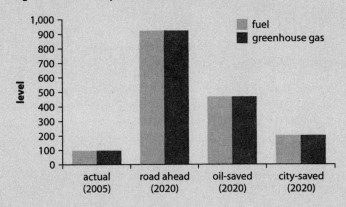

Box Figure. Fuel Consumption and Greenhouse Gas Emissions

Source: World Bank 2006.
Note: The 2005 greenhouse gas emission level = 100.

The simulations of the oil-saved and city-saved scenarios derive from the proven performance of advanced technologies and policy instruments. Provided that there is further technological advancement in the next 15 years, the payoff by 2020 could be even larger. In addition to the savings in energy and reductions in greenhouse gas emissions, the city-saved scenario would encompass other desirable features of sustainability. For example, the design of compact cities with higher population densities, efficient public transportation, and rational fuel pricing would save farmland while maintaining the efficiency of the urban economy.

It is not difficult to see what the policy directions should be for China. The 2004 automotive industrial policy encourages the development of energy-efficient and low-emission vehicles. It also calls for more effort to research and develop low-sulfur vehicular fuels and alternative fuel vehicles such as hybrid electric vehicles. Clearly, the policy is heading in the right direction for improved energy sustainability of the transportation sector, but it could be enhanced by encouraging the development and production of small electric vehicles suitable for low-speed driving in a dense urban environment.

Source: World Bank 2006.

Box 3.3

Energy Efficiency in Buildings

With the French Global Environment Facility: Reducing Energy Consumption in Buildings by 50 Percent

China and France have established a significant cooperative program for improving energy efficiency and making housing affordable. This program, which has existed since 1999, involves the central provincial and municipal authorities. The sponsors are the French ministries in charge of environment and external relations under the aegis of the French Global Environment Facility.

The program's primary objective has been to achieve energy savings in housing. Its activities have focused on housing in three northern provinces—Heilongjiang, Liaoning, and Beijing—where winter temperatures can be severe, dropping below −40°C. It has built about 789,000 square meters of energy-saving housing, reducing energy consumption by 50 percent and preventing the release into the air of about 44,000 tons of carbon dioxide annually—and these savings are available at only 7 percent more than the normal construction costs in China.

The program uses a three-pronged approach to achieving these savings:

- Applying energy-saving building techniques, including insulation, waterproofing, and so on.
- Strengthening technical and industrial partnerships. Chinese investors are studying two French building insulation processes.
- Supporting local institutions in adapting the regulatory framework to provide an incentive system for controlling energy use in housing.

With the World Bank: Reducing Space Heating Energy Consumption in Buildings by 50 Percent

The potential for and difficulties of achieving energy savings in Chinese buildings are exemplified in northern Chinese cities, where most buildings are connected to coal-fired central heating systems. Energy waste in centralized heating is excessive because of the prevalence of poor thermal integrity in buildings, antiquated heat supply technologies, and welfare heat pricing and billing practices. The government reckoned that by enforcing the existing national standards on building energy efficiency, the energy demand for space heating of new residential buildings could be halved.

> The World Bank is working with the government in a comprehensive effort to address the root problems of energy waste in centralized space heating. The China Heat Reform and Building Energy-Efficiency Program, funded by the Global Environment Facility and other donors, focuses on three key aspects of achieving sustained energy-efficiency improvement in the central heating chain (World Bank 2004):
> - Enforcement of building energy-efficiency standards
> - Technical innovations enabling end-use control of heat consumption and demand-driven heat supply
> - Implementation of cost-based heat pricing and consumption-based billing.
>
> Energy savings in centralized space heating could reach 30 mtce per year by 2020, if all new residential buildings and associated heating systems constructed in northern Chinese cities in the next 15 years meet the current energy-efficiency standards. However, the extent to which this technical potential materializes into real coal savings depends critically on the broad-based implementation of cost-based heat pricing and consumption-based billing.
>
> *Source:* French Global Environment Facility.

International experience suggests that energy-efficiency practices in the building industry are driven by (a) serious regulatory efforts in mandating and enforcing building standards for energy efficiency and (b) market conditions that reward such practices. China is lacking in both aspects. Only about 30 percent of the urban residential buildings constructed in cold-climate regions in 2004 complied with current national energy efficiency standards. China's centralized heating services are still deeply entrenched in a welfare-based system that denies end-use conservation and discourages technical innovations: heat metering is virtually nonexistent, heating services are priced and billed by fixed rates based on floor area, and most employers pay at least part of the heat bills of employees.

Rapid urbanization and lack of planning capabilities left China's major cities trying to cope with growth by following past urbanization models rather than by developing more efficient models that are better adapted to the country's specific conditions. The current building stock is estimated at about 39 billion square meters. It is expected to increase to 69 billion square meters in 2020, with new building stock therefore accounting for about 43 percent of the total.

The DRC has estimated that improvements in the energy efficiency of existing buildings and the design of more energy-efficient heating and cooling systems for new buildings could bring about energy savings amounting to 240 mtce, or about 43 to 52 percent of the sector's final energy consumption in 2020. Of those savings, about 60 percent would come from improvements in the design of new heating and cooling systems. In addition, savings in lighting and household appliances would amount to about 94 mtce. Thus, the total energy savings in the residential and commercial sector would be equivalent to about 60 to 70 percent of the sector's projected total energy consumption for 2020.[6] However, without improvements in energy efficiency, China's residential and commercial buildings are likely to consume about 1.1 billion tce in 2020—about three times their consumption level of 2000.

A quadrupling of income levels is likely to increase the demand for household appliances in China. Data show that the penetration of color television sets, refrigerators, washing machines, and electric fans is fairly high. The penetration of air conditioners, one of the more energy-intensive appliances, has grown rapidly from only 8 per 100 households in 1995 to 51 per 100 in 2002 (table 3.5). The major expansion of their use in China contributed to the soaring summer electricity load and blackouts in southern China in the early 2000s. In preparation for substantial increases in appliance use, China has developed one of the most comprehensive appliance standards and labeling programs in the developing world. This program has mandatory minimum standards covering 15 types of appliances, lighting, and industrial products, along with a voluntary endorsement label and a proposed information label. The

Table 3.5 Household Appliances per 100 Households, 1995–2002

Appliance	1995	1996	1997	1998	1999	2000	2001	2002
Air conditioner	8.09	11.61	16.29	20.01	24.48	30.76	35.79	51.10
Washing machine	89.97	90.06	89.12	90.57	91.44	90.52	92.22	92.90
Electric fan	167.97	168.07	165.74	168.37	171.73	167.91	170.74	182.57
Refrigerator	66.22	69.67	72.98	76.08	77.74	80.13	81.87	87.38
Freezer	2.87	3.48	4.46	4.80	5.37	6.52	6.62	6.81
Color television	89.79	93.50	100.48	105.43	111.57	116.56	120.52	126.38
Electric cooker	84.14	91.50	92.35	95.98	101.82	101.94	107.87	—
Shower heater	30.05	34.16	38.94	43.30	45.49	49.11	52.00	62.42

Source: China National Bureau of Statistics various years.
— = not available.

standards are not adequately implemented, however, because of weak enforcement and lack of information and financial incentives.

The Missing Link: An Improved Policy and Institutional Framework

A fundamental refocusing of China's energy-efficiency policies is urgently needed. In the past, the focus of energy efficiency in China was on energy-saving measures during times of shortages. Historical data indicate higher levels of investment in energy efficiency during supply-constrained periods. The share of energy-efficiency investments in China's total energy investments ranged between 8 and 14 percent during the 1980s, a period of severe energy and power shortages, but fell to only 6 to 8 percent in the early 1990s, when supply constraints eased. Since the mid 1990s, policy makers have tended to focus on broader issues, such as oil and gas and power subsector reforms, rather than on the specifics of energy savings.

As rising energy intensity provokes concern about China's energy security, the focus needs to shift to making energy efficiency one of the top national priorities (DRC 2004). Engaging the whole society in this endeavor is essential to its success. The theme of awareness campaigns should not be on catching up to the efficiency levels of industrial countries but instead on developing and rapidly deploying innovative, cost-effective solutions adapted to the size and composition of China's energy market and matching the challenges facing it.

Technological leapfrogging is extremely important, but effective institutions and an adequate incentive system matter even more. There are many advanced technologies for energy efficiency that other countries have developed but have not applied because of the lack of significant market size and cost-effectiveness. China could benefit substantially by acquiring and developing these technologies and launching research and development (R&D) programs to advance them. In this way, the country could become a world leader in this field, preparing for the next stage of economic development by concentrating on resource-use efficiencies, after the current high-growth era driven mainly by manufacturing. China's large market and demonstrated ability for reducing technology costs by producing equipment domestically could make these technologies cost-effective, not only for China but for other countries as well.

In the early stages of its development, Japan purchased the licenses to a number of advanced technologies and adapted them to develop one of the most energy-efficient economies in the world. In following this example, China also would have to devote a much larger budget to R&D, with a particular focus on energy efficiency. Currently, the country spends only about 1.8 percent of what Japan spends on energy R&D. In addition, the government would have to provide incentives for the private sector to invest in energy efficiency. Currently, R&D expenditures by private enterprises for energy-saving technology account for about 2 percent of all expenditures on energy-related R&D.

Since 1997, China has had an energy conservation law. In addition, there are about 164 state standards for energy savings, technology demonstration projects, and policies for demand-side management. But serious problems continue to plague the implementation and enforcement of standards. There are four principal reasons for the disconnect between what the law and policies are meant to do and what the actual outcomes are:

- The restructuring and the opening and decentralization of the economy weakened the command-and-control institutional framework and system, which had been very successful at reining in energy consumption between 1980 and 2000, but failed to replace it with an institutional framework and policies that would have the same effect in a more market-oriented economy.
- There is insufficient government capacity to enforce existing laws and standards at the national level and particularly at lower administrative levels.
- The energy shortages of the mid 1980s resulted in an exclusive focus on supply additions and therefore ignored lower-cost energy-efficiency options, indicating serious flaws in the planning system.
- Perhaps most important, in the absence of a comprehensive reform, the pricing system became less and less suited to the rapidly changing energy environment. Even worse, the piecemeal pricing changes introduced distortions that led to suboptimal use of resources and energy waste.

The high priority that the proposals for the 11th Five-Year Plan have accorded to energy efficiency is very encouraging. However, given the closing window of opportunity, it is critical to translate the sound general principles into concrete action plans. The first step in this process should be to address the disconnect between the intent of the law and regulations for energy efficiency and the actual outcomes. More detailed recommendations are provided in chapter 7.

Notes

1. For example, the energy intensity of the U.S. economy climbed for about 70 years, reaching its peak in 1920. The United Kingdom reached its peak energy intensity in about 40 years but at a higher level than in the United States. In both countries the energy-intensity path to the peak was very steep. In contrast, Japan's rise to its peak energy intensity took place more gradually over a 60-year period. However, Japan's peak energy intensity, at about 0.2 tons of oil equivalent (toe) per US$1,000 of GDP, was dramatically lower than that of the United States (about 0.9 toe per US$1,000 of GDP) or the United Kingdom (1.05 toe per US$1,000 of GDP).
2. A steep rise in energy intensity during the Great Leap Forward of the 1950s reflected a period when decentralized, inefficient, and energy-intensive small-scale industries were driving the economy. A sharp decline in energy intensity occurred in the early 1960s, mirroring the time known as the Years of Great Difficulty, characterized by natural disasters, poor harvests, and industrial decline. A resurgence of higher energy intensity followed during the Cultural Revolution, as China turned inward for economic development through the 1970s, using inefficient production processes characterized by high levels of waste.
3. A recent World Bank (2006) study estimates that car ownership alone could reach 170 million by 2020 under a business-as-usual scenario.
4. Staggered implementation has been adopted in the United States (a) to accommodate the auto industry, which required time to adapt its installed manufacturing capacity to the new standards; (b) to avoid stranded costs; and (c) because of the lobbying powers of concerned parties. Stranded costs either do not exist theoretically in China or are very low, and vested interests have not developed yet. Brief information on the U.S. corporate average fuel economy standards is available at http://www.dieselnet.com/standards/us/fe.php.
5. This study observes that urban households' mobility needs have been underserved "by the established practice that concentrates much of the available resources to meeting the demand for auto-mobility" (World Bank 2006, 3).
6. This estimate is based on the savings estimated in table 3.5 and the sector breakdown of projected final demand in 2020 by the DRC, amounting to 555 mtce for the ordinary efforts scenario and 463 mtce for the green growth scenario.

References

Berrah, Noureddine. 1983. "Energie et Développement." *Revue de l'Énergie* 356 (August–September): 409–15.

BP Global. 2004. *Statistical Review of World Energy 2004*. London: BP Global.

Chen, Yong, Ugo Farinelli, and Thomas B. Johansson. 2004. "Technological Leapfrogging: A Strategic Pathway to Modernization of the Chinese Iron and Steel Industry?" *Energy for Sustainable Development* 8 (2): 18–26.

China National Bureau of Statistics. Various years. *China Statistical Yearbook*. Beijing: China Statistics Press.

Communist Party of China Central Committee on Drafting the 11th Five-Year Plan. 2005. *The Proposals of the CPC Central Committee on Drafting the 11th Five-Year Plan*. Beijing: Communist Party of China.

DRC (Development Research Center). 2004. *Basic Conception of the National Energy Strategies*. Beijing: DRC.

ERI (Energy Research Institute). 2001. *China Energy Development Report*. Beijing: ERI.

Fisher-Vander, Karen, Gary H. Jefferson, Hongmei Liu, and Quan Tau. 2004. "What Is Driving China's Decline in Energy Intensity?" *Resource and Energy Economics* 26: 77–79.

French Global Environment Facility. Accessible at http://www.ademe.fr/anglais/publication/pdf/va_chine.pdf.

Gu, Zhaoguang, David Moskovitz, and Jianping Zhao. 2005. *Demand-Side Management in China's Restructured Power Industry: How Regulation and Policy Can Deliver Demand-Side Management Benefits to a Growing Economy and a Changing Power System*. Washington, DC: World Bank.

NDRC (National Development and Reform Commission). 2004. *China Medium- and Long-Term Energy Conservation Plan*. Beijing: NDRC.

World Bank. 2001. *China: Opportunities to Improve Energy Efficiency in Buildings*. Washington, DC: World Bank.

———. 2004. *China: Heat Reform and Building Energy-Efficiency Project*. Project Appraisal Document. Washington, DC: World Bank.

———. 2006. "China: Building Institutions for Sustainable Urban Transport." EASTR Working Paper 4, World Bank, Washington, DC.

CHAPTER 4

Greening the Energy Sector

> *Key Messages*
>
> *The projections of atmospheric pollution from energy sources in 2020 show that even in the lowest green growth case, energy consumption will result in pollution levels that exceed the government's proposed environmental caps and significantly contribute to the growth of global carbon dioxide emissions over the period through 2020. China's per capita energy consumption is still low by international standards. However, given the enormous size of the energy sector, emissions of sulfur dioxide and nitrogen oxides are serious threats to the environment and quality of life of the population. The projections by the Development Research Center (DRC) and the Energy Research Institute (ERI) show that the situation would substantially worsen in the future. Even under the green growth scenario, the levels of sulfur dioxide and nitrogen oxides emissions would be 100 percent and 37 percent higher, respectively, than the absorption capacity of the environment.[1] If China were to follow the development path of the industrial market economies (an energy/gross domestic product elasticity greater than 1.0 instead of its historical 0.5 used in all forecasts), the levels of these pollutants would be substantially higher—exceeding by far the pollutant caps envisaged by the government and seriously threatening the sustainable development of the energy sector.*

Furthermore, though China's per capita emissions of carbon dioxide are still lower than the world average (2.7 tons compared with 3.9 tons), projections indicate that China will be responsible for one-fourth of the increase in global emissions in the next two decades.

The creation of a path to greener development in an economy that depends on coal will require an innovative program of actions to reconcile energy supply security with environmental protection and to strengthen international cooperation, aimed particularly at implementing the most advanced energy technologies. China's abundant, low-cost domestic coal has helped to hold down the country's dependence on foreign energy supplies, thereby making the energy system less vulnerable to supply disruption and sudden price hikes. Coal must continue to play a major role in meeting the country's surging energy needs for the foreseeable future. China needs, therefore, to rely extensively on the most advanced efficient and clean-coal technologies (as it is doing with nuclear energy, whose contribution is dwarfed by coal). These technologies must be combined with strong policy incentives to minimize environmental damage. Given the complexity of China's energy sector and the lead times required for change, a sustainable program to ensure environmental protection without limiting economic growth and lessening the security of energy supply will require a coordinated set of actions on several fronts: integration of environmental and energy policies, improved enforcement of existing policies, stricter policies governing emissions controls, increased efficiency of energy use, urgent promotion of clean-coal technologies, and substitution of cleaner fuels for coal.

China already has entered into some bilateral cooperative arrangements for the development of advanced technologies. However, these initiatives are modest compared with the nation's needs. Therefore, in addition to establishing policies that provide greater incentives for research and development of clean-coal technologies, China should continue to expand bilateral cooperation and take advantage of multilateral cooperative partnerships such as the Clean Development Mechanism (CDM). Studies of the potential use of the CDM in China have indicated that the country could capture about 50 percent of the global CDM market

(World Bank and others 2004). Channels to access the most advanced energy efficiency, clean-coal technologies, and carbon sequestration are essential to ensuring sustainable development of the energy sector.

Pollution Levels Still a Major Concern

The establishment of a sustainable energy sector that depends on coal must resolve inherent tensions between securing energy supply and protecting the environment. China's abundant, low-cost domestic coal has helped the country reduce its dependence on foreign energy supplies, thereby making the energy system less vulnerable to supply disruption and sudden price hikes. At the same time, China has suffered considerable environmental damage along the path to economic growth.

The consumption of energy—mainly coal—is the principal source of anthropogenic air pollution in China. Coal accounts for an estimated 70 percent of carbon dioxide emissions, 90 percent of sulfur dioxide emissions, and 67 percent of nitrogen oxide emissions (DRC 2000). It is also responsible for virtually all soot emissions, which represent more than half of the country's pollution from total suspended particulates (TSP). In addition, emissions from the indoor use of solid fuels for cooking and space heating are damaging, though there has been no national assessment of their severity. However, studies have shown that indoor air pollution levels can be several orders of magnitude higher than the U.S. Environmental Protection Agency's National Ambient Air Quality Standards for urban areas (World Bank 2001).[2]

Energy-related air pollution is a major health and ecological issue in China. Studies have shown a link between the high incidence of premature death in the country and both outdoor and indoor air pollution caused mainly by pervasive coal use. Acid rain from increased sulfur dioxide concentrations in the atmosphere—which can retard forest and crop growth as well as endanger aquatic life—affects about a third of China's total land area. A World Bank study made a conservative estimate that air and water pollution combined resulted in losses of about 8 percent of the country's gross domestic product (GDP) (World Bank 1998).[3] This estimate included health losses caused by urban and indoor air pollution, chronic disease from water pollution, and crop and forestry damage from acid rain. In terms of magnitude, this share represents more than twice the share of China's GDP spent on education (Qiao 2005).[4] More recent studies indicate even greater economic impacts.

The main sources of air pollution from energy use are large, stationary sources such as power plants and heavy industry. The electric power industry and three other industries account for 65 percent of sulfur dioxide emissions and 69 percent of TSP emissions from energy sources (Ho and Jorgensen 2003).[5] Coal use in power generation is by far the largest single industrial source of pollution, accounting for 45 percent of sulfur dioxide emissions and 35 percent of TSP emissions. These data suggest that targeting technologies and policies to reduce emissions in a relatively small number of large stationary sources could substantially reduce overall emissions (World Bank 2001, 93, 117).

In addition to large stationary sources, the rapid expansion of the motor vehicle fleet in the major cities in recent years has augmented air pollution from fossil fuels. For example, in a few very large cities—including Beijing, Shanghai, and Guangzhou—mobile sources of pollution account for 45 to 60 percent of nitrogen oxide emissions and 85 percent of carbon emissions (World Bank 2001). Motor vehicle penetration in China is fairly low compared with other countries. However, projections indicate more than a sevenfold increase in the vehicle fleet between 2000 and 2020. To prevent these potentially large, decentralized sources of pollution from growing uncontrollably, China has a closing window of opportunity to develop urbanization strategies that reduce mobility needs and transportation strategies that are less dependent on privately owned cars. Strict enforcement of existing standards could also significantly contribute to restraining the growth of emissions.[6]

Despite overall aggregate reductions of pollution, air quality varies substantially by area and city size. The 1980s and 1990s saw these main trends in emissions of the major pollutants:

- A substantial reduction in sulfur dioxide
- Slight reductions in TSP in aggregate
- A small increase in nitrogen oxide levels, reflecting the growing impact of transportation vehicles on overall air quality
- A large increase in carbon dioxide levels.

Trends varied significantly by city size during this period. For example, although TSP registered an overall decline, its concentration actually increased by a factor of two or more in a few large metropolitan areas, such as Beijing and Shanghai. In small cities, concentrations of all three major pollutants increased. One reason for this increase is that small cities tend to rely more on coal than on cleaner fuels like natural gas to meet the energy needs of their residential and commercial sectors (World Bank 2001).[7]

In its report on the state of the environment for 2004, China's State Environmental Protection Administration (SEPA) noted that about 61.4 percent of the 340 cities that SEPA monitors did not meet air quality standards for urban areas (SEPA 2003). Particulates are the main pollutant affecting urban air quality, with concentrations higher than the national standard for urban areas in about 54 percent of cities. Some 64 of the 342 cities in sulfur dioxide control zones did not meet the national standard. The cities with the most serious pollution from sulfur dioxide pollution were located in the provinces of Shanxi, Hebei, Henan, Hunan, Inner Mongolia, Shaanxi, Gansu, Guizhou, Chonquing, and Sichuan. Of the 116 cities in the acid rain control zones, 74 percent met minimum urban air quality standards (SEPA 2003).

The primary global environmental concern for the development of China's energy sector is the size and rate of growth of carbon dioxide emissions. This is true despite China's low per capita emissions of 2.7 tons, compared with a world average of 3.9 tons and a U.S. figure of 20.1 tons. However, because of China's massive population, the country's aggregate carbon dioxide emissions amount to nearly 14 percent of the global total, second to the United States, which has the highest share at 24 percent. However, China's share of carbon dioxide emissions is still less than its share of the world's population (21 percent).[8]

Forecasts by the International Energy Agency (IEA) project that China will be responsible for one-fourth of the world's growth in carbon dioxide emissions between 2002 and 2020 (figure 4.1, see p. 72). These projections also show that by 2015 the country will begin to produce more of these emissions than all the European countries of the Organisation for Economic Co-operation and Development (OECD) combined. China has signed the Kyoto Protocol to the United Nations Framework Convention on Climate Change (UNFCCC), but as a developing country, it is not subject to emissions targets. Nevertheless, China is concerned about curbing increases in these emissions and is seeking partnerships to develop technologies that will help cut them.

As a party to the UNFCCC, China has submitted a report titled *The People's Republic of China Initial National Communication on Climate Change* (Government of China 2004). In the report, China recognizes that parties to the Convention should "on the basis of equity and in accordance with their common but differentiated responsibilities and respective

Figure 4.1 Expected Growth in Global Carbon Dioxide Emissions through 2020

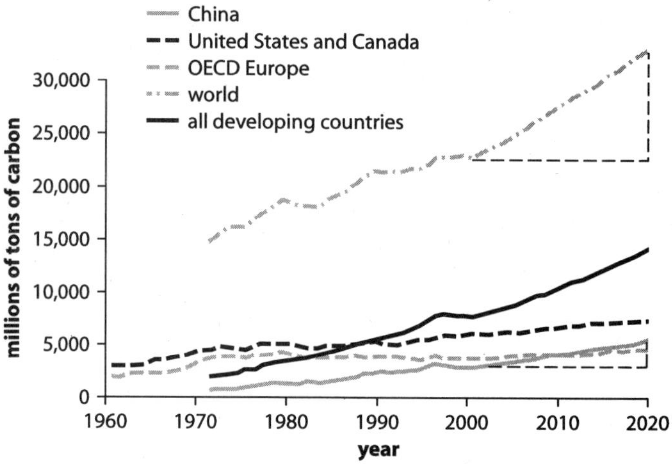

```
Legend:
— China
– – United States and Canada
– – – OECD Europe
· – · world
—— all developing countries
```

y-axis: millions of tons of carbon (0 to 30,000)
x-axis: year (1960 to 2020)

Source: IEA 2004.

capabilities, protect the climate system for the benefit of present and future generations of mankind" (Government of China 2004, iii). China also presented its national inventory of greenhouse gases and noted its needs for funds, technologies, and capacity building in order to fulfill its obligations under the UNFCCC. In this context, China outlined its need for "technologies related to environmental protection and the comprehensive utilization of resources, various energy technologies, advanced technologies for transportation, advanced technologies related to material and manufacturing industries, building sector technologies, etc." (Government of China 2004, 16).

Environmental Impacts of the Energy Scenarios

China's economy will still be highly dependent on coal in 2020. It will require an integrated program of technology leapfrogging and integrated energy and environmental policies to safeguard future generations.

The DRC and ERI energy scenarios show that coal will continue to dominate energy supply in 2020, with a share ranging from 59 percent to 63 percent. However, the energy consumption scenarios and their related pollution impacts are differentiated on the basis of assumptions about policies for environmental protection to promote energy sustainability by 2020 (see table 2.3). These assumptions are (a) improved enforce-

> **Box 4.1**
>
> **Key Elements for Environmental Improvement in the Green Growth Scenario**
>
> The green growth scenario suggests these elements for achieving environmental improvement:
> - *Environmental standards and enforcement.* Create a more stringent legal system.
> - *New control policy.* Tighten emissions standards in large cities.
> - *Energy-efficiency improvements.* Keep energy efficiency of technology sectors on track to reach current advanced international levels (see chapter 3) by 2030.
> - *Clean-coal technology.* Achieve these goals: (a) 50 percent of coal-fired units to have desulfurization units by 2020, (b) clean-coal technology put into operation for demonstration purposes by 2010, and (c) integrated coal gasification combined cycle power generation units commercialized by 2015 with 45 percent efficiency.
> - *Fuel substitution.* Increase substitution of gas in medium and large cities and generate 11 percent of power from gas-fired capacity.
>
> *Source:* DRC 2004.

ment of existing policies, (b) stricter policies governing emissions control, (c) increased efficiency of energy use, (d) promotion of clean-coal technologies, and (e) substitution of cleaner fuels for coal.

All three scenarios (ordinary efforts, promoting sustainability, and green growth) assume that by 2010 the government of China will achieve current environmental standards for sulfur dioxide emissions. Other than meeting those standards, the ordinary efforts scenario assumes no additional efforts in the energy sector to support environmental projection between 2010 and 2020. The promoting sustainability scenario includes a stronger emphasis on standards to control the emissions of fine particulates and some substitution of gas for coal in large cities, but no emphasis on expanding clean-coal technologies. Only in the green growth scenario are major efforts assumed to be made to reduce the environmental impact of China's energy consumption on all five fronts, as box 4.1 indicates.

The DRC has analyzed the levels of key air pollutants associated with the ERI's energy scenarios compared with the caps that the government is considering for these pollutants (table 4.1). The analysis shows that China already faces an environmental deficit for sulfur dioxide, with atmospheric concentrations exceeding environmental absorption capacity.

Table 4.1 Projections of Key Air Pollutants in the Energy Scenarios for 2000–20 and Government Caps
million tons

Indicator	Actual figure for 2000	Projections for 2010			Projections for 2020		
		Low	Medium	High	Low	Medium	High
Sulfur dioxide emissions	26.5	26.8	30.4	31.7	27.9	34.0	39.4
Environmental cap	20.0	16.0	16.0	16.0	13.0	13.0	13.0
Nitrogen oxide emissions	18.8	24.6	27.3	28.5	28.7	35.0	40.6
Environmental cap	19.0	18.0	18.0	18.0	16.0	16.0	16.0
Smog and dust emissions	125.6	184.9	209.5	218.7	233.5	284.8	330.3
Carbon emissions	831.6	1,073.0	1,215.0	1,269.0	1,259.0	1,535.0	1,781.0

Source: DRC 2000.

In addition, levels of nitrogen oxides are almost at environmental capacity. The projections show that the situation would worsen substantially in the future. Even under the green growth scenario, the levels of sulfur dioxide and nitrogen oxides would be 100 percent and 37 percent, respectively, higher than the environment's absorption capacity.[9] If China were to follow the development path of the industrial market economies (an energy/GDP elasticity greater than 1.0 instead of the projected 0.5), the levels of these pollutants could be even higher than those in the ordinary efforts scenario, which are triple and double the caps on sulfur dioxide and nitrogen oxides. Therefore, it is important for China to address the apparent disconnect between high environmental awareness at the government level, as reflected in the proposed pollution caps and advanced environmental policies, and investment strategies that still rely on high-polluting technologies without regard to the costs of externalities. As shown in the green growth scenario, meeting the caps would require early and rapid deployment of advanced technologies for higher levels of efficiency and cleaner use of coal.

The Energy Path to Greener Development

China can avoid a pollution-intensive path through a combination of technological leapfrogging and innovative policies for cleaner energy development. The country is now facing the difficult challenge of finding ways to green its economic development while continuing to use its abundant domestic coal as the main energy source. Weaving environmental concerns into the fabric of industrialization at such an early stage is unprecedented. There are no off-the-shelf models from the industrial market economies that China can apply directly. However, numerous studies have shown that China could reconcile environmental concerns with economic development needs by placing these concerns at the core of its development policy and by leapfrogging to cutting-edge energy efficiency and clean-coal technologies, as it is doing with nuclear energy.[10] This technological leapfrogging could provide the country with the opportunity to become a leader in the business of green development rather than a follower of the "pollute now, pay later" model. However, achieving this position will require prompt actions to establish incentive policies for making the shift to a new development model before the growing economy becomes locked into technologies that are more pollution intensive.

The development model for the industrial market economies in the past has been a "pollute now, pay later" path to economic growth and higher per capita income levels. To illustrate this path, economists often use an environmental Kuznets curve, which shows a path of rising pollution levels to attain higher per capita income. Environmental concern became an afterthought of the industrialization process, with the emphasis on cleanup operations and improved technology to remove the damage that the early phase of rapid economic development had caused.

Although recent studies have raised questions about the econometric robustness of the model and its validity for all pollutants, the model still reflects the approach that most developing countries take in addressing environmental issues (Stern 2003). The relationship the curve describes is of particular concern to China because the economic basis for the country's energy projections between 2000 and 2020 is an increase in per capita income from less than US$1,000 to nearly US$4,000. This concern is certainly understandable, given that the DRC and ERI's energy scenarios for 2020 forecast, even in the most environmentally optimistic cases, that all pollutants are likely to exceed their maximum safe concentrations.

An expanded legal framework and the recent proposals for the 11th Five-Year Plan (2006–10) have given substantial impetus to creating a new development path to reconcile both security of supply and protection of the environment. China's past laws on air pollution were general and established limited avenues for addressing infractions. The enactment of China's Air Pollution and Prevention Control Law in 2000 was therefore a major step toward a greener development path, reflecting the growing concerns of China's policy makers about the high costs of pollution to the economy. The law constitutes a major departure in both its substance and in the details of its application. It calls for measuring the total atmospheric load of a pollutant, elaborates legal responsibilities in more detail, and specifies fines for noncompliance. Perhaps most significantly for China, it opens the door for greater involvement of the public in environmental policy by requiring medium-size and large cities to release regular reports detailing the types and effects of local pollutants and informing the public about environmental hazards. In addition, China has developed a vast array of pollution control regulations at the national, provincial, and local levels (Alford 2002).

Building on the supportive legal framework, the proposals for the 11th Five-Year Plan have called on China to "make more efforts on environmental protection. Instead of old practices that pollute first and treat next, we shall try to stop pollution at the very beginning. Environmental protection is a priority of departments and governments in various regions. Strict measures shall be taken to reduce the overall volume of emission" (Communist Party of China Central Committee on Drafting the 11th Five-Year Plan 2005, 12). However, creating policies that will give effect to these intentions and laws for the country's fast-moving economy, which has become increasingly decentralized and more market oriented, will require strong central and local government coordination combined with a comprehensive plan of action and the institutional capacity to implement it. In this regard, the DRC has emphasized four key aspects to take into account in implementing an effective environmental strategy:

1. Implementation of a system of market incentives to accelerate the development of clean energy
2. Establishment of stricter caps on emissions in order to maintain the balanced development of the economy, energy use, and the environment
3. Promotion of emissions trading, given the impressive results achieved in other countries
4. Increase in the public awareness of environment problems and the participation of nongovernmental organizations in the development process, to ensure that the government executes its executive authority fairly and in accordance with the law.

The legal framework for the environment and the main principles for the 11th Five-Year Plan require translation into more specific policies and plans of actions. As policy makers design an action plan, it would be helpful to produce an energy development scenario that evaluates recent growth trends and is consistent with environmentally sustainable levels of the major air pollutants, because even the green growth scenario would fail to meet the proposed caps on major pollutants. The new scenario would have to be more supportive of the clean-fuel alternatives (such as gas, renewable energy, and nuclear energy with strict and enforced safety and security standards) and the advanced clean-coal technologies that are vital to balance economic growth with energy needs and environmental sustainability. This analysis should focus particularly on which technologies are likely to bring the greatest environmental gains for China.

If environmental concerns are to become the basis for making decisions on investments in technologies for energy use and production, it is important to determine baselines for environmentally acceptable limits on pollutant concentrations. Given the baselines, the share of pollution that specific energy-consuming activities contribute would be assessed. On the basis of the acceptable levels, it would be possible to determine, for each sector, the most appropriate technologies for energy efficiency and pollution control, both of which will contribute to reducing emissions. Doing so will enable the country to meet those emissions caps or, if not, at least determine the environmental damage costs of exceeding them.

Decision makers need to be aware of the costs of high pollution levels in order to make investment decisions about technologies for improved energy efficiency and reduced emissions. Preparing alternative energy consumption scenarios through 2020 is a first step toward a less energy-intensive and cleaner path of industrialization. To steer the country toward this more sustainable path, decision makers will need to know the costs and benefits that it will entail. They will need to analyze those costs and benefits over the full life of the assets required to achieve sustainable development by 2020. However, long-term energy strategies should extend further, preferably to 2050, to take account of the potential effects beyond 2020 of using more advanced technologies such as carbon sequestration and hydrogen-based fuel cells. Such technologies could become economical in the future, especially with China and, potentially, India as market drivers.

Including environmental costs in investment analyses can make a dramatic difference in decision making about new technologies. For example, as stressed in chapter 3, integrated coal gasification combined cycle

(IGCC) technology for power generation has been characterized as too expensive when only its capital cost—which is nearly double that of conventional technology—was considered. However, very simple analysis taking into account a conservative estimate of environmental costs shows that with only a 15 percent reduction in capital cost, the technology could become the most economical choice for power generation in the future (see appendix F). Such a reduction is well within the reach of Chinese industry.[11] Furthermore, this technology, in addition to using less coal per unit of output, would allow for easier carbon sequestration in the future, should the latter become economical. All commercially available technologies that are not deployed in industrial countries because of market constraints or stranded costs or vested interests should be evaluated. Assessments should take into account the technologies' economic viability, conditions for access, and scope for reducing costs in China. This exercise should be followed with aggressive implementation strategies for early deployment of any technologies that could significantly contribute to putting the energy sector on a sustainable path.

Energy modeling has shown that China could meet its long-term energy needs using advanced technology to produce clean fuels derived mostly from indigenous resources. Researchers have experimented with long-term planning models for China, but the work is several years old and needs updating. For example, researchers from China and the United States jointly prepared a model to evaluate whether an advanced technology strategy could possibly be cost-effective for China's energy sector development between 1995 and 2050. In July 2001, they published a study on the subject for the Working Group on Energy Strategies and Technologies (WGEST) of the China Council for International Cooperation on Environment and Development (Wu and others 2001). The study was the product of nine years of research on many advanced energy technologies of long-term strategic interest to China. Technologies that were examined included coal gasification, coal-based electricity production with carbon dioxide sequestration, and hydrogen fuel cells.

The researchers preparing the WGEST study used the MARKAL model, an energy planning tool developed from the late 1970s.[12] The application of the model first defines geographic boundaries and then builds a representation of an energy system specifying material and energy flows in and out of each technical component of the system. For each technology, the model includes economic data, performance data, and emissions output. The MARKAL model for China was based on work at Tsinghua University. The assessment of Wu and others (2001) expanded an earlier MARKAL model to include a set of advanced technology options and their effects on China's energy demand, supply, and emissions during the period through 2050.

The results of the study showed that leapfrogging to an advanced technology scenario did not involve significantly higher discounted costs than the business-as-usual scenario. In fact, the average capital investments for this scenario were only about 5 percent higher than the business-as-usual scenario through 2035 but increased to about 25 to 35 percent above the base case between 2035 and 2050. However, overall the advanced technology had a lower cost because its higher efficiencies and lower fuel costs more than offset the capital costs over the total period covered. The analysis used a 10 percent discount rate, which is still an indication of a strong preference for the present. Many economists advocate lower discount rates for investments that preserve the environment because of the long-term impact of environmental change. The appropriate discount rate is still a matter of debate among economists and outside the scope of this report. However, it is a worthwhile subject for research by the government: the discount rate is a very important decision-making tool, and traditionally China has been a society that places a greater value on savings for the future than do the industrial market economies.

Using the MARKAL model, Wu and others (2001) concluded that China could support its social and economic objectives for the long term using clean, renewable energy derived mostly from its indigenous resources. In addition to substantially reducing environmental damage, the reduced reliance on foreign energy sources would enhance China's supply security. Another very interesting result was that nuclear energy became an important factor for achieving carbon dioxide reductions in the base case, using existing technologies, but was not necessary at all in the advanced technology case. This result occurred because the use of China's vast coal resources, combined with carbon dioxide sequestration technologies, made it possible for China to meet electricity needs while achieving dramatic reductions in carbon dioxide emissions.

International cooperation in the development of new environmentally sustainable energy technologies and the use of the CDM presents an unprecedented opportunity to safeguard the environment and achieve sustainability. China is engaged in important cooperative ventures to develop advanced energy technologies, which could be beneficial to both China and the industrial countries. China entered into cooperative ventures with the U.S. Department of Energy and the Japanese Ministry of International Trade and Industry to promote advanced clean technologies.

In addition, China signed a declaration with the European Union in September 2005 on cooperation in the field of advanced zero-energy emissions coal technology through carbon capture. Furthermore, in July 2005, at the Gleneagles Summit, chaired by British Prime Minister Tony Blair, leaders of the Group of Eight countries agreed to a plan of action for cleaner fossil fuels, with particular emphasis on advanced coal-use technology. This cooperative trend could help China move more rapidly along the path of attaining its economic aspirations while limiting pollution and thereby enhancing the country's overall energy security.

In addition to controlling pollution in the power subsector, China now has a major opportunity to control pollution in the urban transportation sector, which is in the very early stages of its development—especially the passenger car market. The DRC cites a wide range of policies that should receive attention, including giving priority to public transportation on the basis of clean fuels, tightening emissions standards for passenger cars, improving vehicle maintenance, and promoting alternative fuels. In addition, China's fast-growing market for vehicles could be a catalyst to significantly increase the penetration of advanced technologies, such as hybrid cars, in countries where vehicle markets are smaller and expanding more slowly.

The bilateral and multilateral cooperative arrangements for the development of advanced technologies that are now under discussion represent a significant step in the right direction. However, they are modest compared with China's needs and the global challenges of climate change. Support for the value of international technology cooperation has just come from a major new report (N. Stern 2006) to the British government on the economics of climate change. It states, "Coherent, urgent, and broadly based action [on energy research and development] requires international understanding and co-operation, embodied in a range of formal multilateral and informal arrangements" (N. Stern 2006, 516). The report claims that cost reductions can be boosted "by increasing the scale of new markets across borders" (N. Stern 2006, 516). However, it is less specific about how this can be achieved than might have been expected in light of the British initiative at the Gleneagles Summit.

In addition to pursuing incentive policies for research and development of clean-coal technologies, China should continue to expand bilateral cooperation and take advantage of international cooperative partnerships such as the CDM. Studies of the potential use of the CDM in China have indicated that the country could capture about 50 percent of the global CDM market (World Bank and others 2004). Novel, mutually beneficial cooperative approaches need to be designed and implemented to allow countries with sizable markets, such as China and India, to access available technologies and deploy them on a large scale as soon as possible.

The formulation and implementation of the necessary environmental policies would require a massive effort to expand China's institutional capability for environmental monitoring and planning. Even with the current level of effort, SEPA can achieve only about 10 percent of its stated objectives because of a very limited staff (compared with the 18,000 people working at the U.S. Environmental Protection Agency). But increasing capability will require not only adequate staffing of the environmental institutions at the national and lower levels, but also (a) more focused regulations and processes; (b) fines commensurate with the damage caused by violators; (c) strict (nonnegotiable) enforcement; and (d) transparency, accountability, and increased involvement of civil society.

The use of resources from the highly profitable CDM projects—such as the capture and distribution of hydrofluorocarbons, particularly trifluoromethane, with its very long atmospheric life span (see box 4.2, on pp. 82–83)—should focus on technologies that address long-term issues, such as acquisition of the most advanced energy efficiency and clean-coal technologies. Providing capital subsidies or operating support to, say, a wind farm will have little long-term effect. As an example, a capital grant of US$100 per kilowatt would support an additional 2,000 megawatts of capacity and a subsidy of Y 0.04 per kilowatt-hour produced by the wind farm—the equivalent of the proposed certified emissions reduction (CER) price for Huitingxile[13]—would support about 845 megawatts of wind capacity use. Compared with the existing installed capacity of 750 to 800 megawatts in China, these subsidization schemes would not lead to a major scale-up. Support for the acquisition and adaptation to local conditions of low-carbon technologies, as for conventional technologies, would reduce costs significantly and lead to their scale-up. This move has by far the best comparative advantage for addressing in a sustainable manner long-term environmental concerns, at both the local and the global levels.

The longer-term perspective tends to favor using CDM funding for clean coal and renewable energy sources, because energy efficiency, liquefied petroleum gas, and coal-bed methane are all economical—or near enough to being economical that sales of CERs can make them so. Energy efficiency in some areas—for example, highly efficient buildings in the government sector—could also be considered. Similarly, support for demonstrations of new technologies that are the first of a kind might be warranted, if there is substantial potential for replication.

Box 4.2

China's Clean Development Fund: Leveraging Carbon Finance for Technology Transfer

As a developing country that approved the Kyoto Protocol in August 2002, China is eligible to participate in projects for selling certified emission reductions (CERs) under the CDM. China now has a wide range of CDM projects under implementation. The provisions of article 24 of the national "Measures for Operation and Management of Clean Development Mechanism Projects in China" allow a unique opportunity to leverage a share of revenues from CER sales toward investments in transfer of much-needed technology related to cleaner energy production in China. Under article 24, the government has specified joint ownership of revenues generated from CER sales and will receive 65 percent of revenues from all hydrofluorocarbon and perfluorocarbon projects, 30 percent from all nitrogen dioxide projects, and 2 percent of revenues from certain "priority sectors." The revenues will be used to support activities related to climate change.

On December 19, 2005, the World Bank's Umbrella Carbon Facility signed emission reduction purchase agreements for contracts totaling €775 million (US$930.9 million) with two private chemical companies: Jiangsu Meilan Chemical Group and Changshu 3F Zhonghao New Chemicals Material Co. The Umbrella Carbon Facility will purchase the CERs from the capture and destruction of trifluoromethane, one of the most potent greenhouse gases, a hydrofluorocarbon with a global warming potential that is 11,700 times that of carbon dioxide. Trifluoromethane is generated as a waste gas in the manufacturing process of chlorodifluoromethane, a hydrochlorofluorocarbon that is used as a refrigerant and as a feedstock, a raw material for other products. Hydrofluorocarbons are among the six greenhouse gases covered under the Kyoto Protocol.

The World Bank is also working with China's Ministry of Finance to establish a Clean Development Fund to receive and manage the government's share of the CER revenues. The fund will apply to all future projects in China. It is considered a model for revenue-sharing arrangements in other countries. The World Bank is providing support to the government in the design of the fund, including areas to be covered, project selection criteria, institutional arrangements, and operating rules. The stated objective of the fund is to provide financial support for the design and implementation of projects and activities in the areas of climate change mitigation and adaptation, with priority focus on energy efficiency and renewable energy.

The first two agreements—those with Jiangsu Meilan Chemical Group and Changshu 3F Zhonghao New Chemicals Material Co.—will eventually generate up to €503.75 million (or US$604.5 million) of fund revenues by 2013. Additional projects coming on line in the future will significantly expand the size of the fund and provide an unprecedented opportunity to leverage resources toward the above priorities, including the costs of transferring technologies to achieve the energy sustainability that will be crucial to China's development objectives over the long term.

Source: Authors.

Figure 4.2 (on p. 84) illustrates the approach by highlighting areas that could be considered because of their potential for helping achieve sustainability. The main focus would be on piloting new applications (such as offshore wind or IGCC demonstration, especially with carbon dioxide sequestration) and acquiring licenses for advanced technologies. A fund for helping Chinese companies pay license fees could be considered. The aim would be for China, which has the largest installed base of coal-fired capacity in the world, to improve its expertise so that it can become a leader in these areas. The same aim could be applied for India. This approach will require an international cooperative strategy, well beyond the current initiatives.

Notes

1. The DRC and ERI's data show no caps on smog and dust or carbon emissions. They show caps on smog only, but there is no disaggregation of smog and dust data for comparison.
2. Studies in the late 1990s (Murray and Lopez 1996; World Bank 1997) estimated that air pollution could be responsible for as many as 700,000 premature deaths in China per year.
3. This estimate is based on a willingness-to-pay evaluation. The direct health and productivity losses (valued in terms of the costs of lost workdays, hospital and emergency room visits, and the debilitating

Figure 4.2 Funding the Technological Leapfrogging

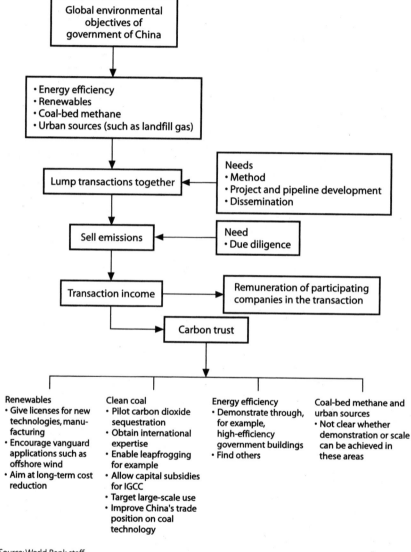

Source: World Bank staff.

effects of chronic bronchitis) were estimated at US$20 billion, representing the single largest environmental cost in China.
4. Qiao (2005) reports China's spending on education as being "no higher than 3 percent of GDP."

5. Ho and Jorgensen's (2003) study was based on data for 1997–99.
6. Industrial countries have shown that total pollution emissions from transportation vehicles can be reduced by 30 to 40 percent by good service and maintenance.
7. In larger cities, a greater share of the residential and commercial sector has substituted gas for coal. For example, between 1991 and 1998, the number of urban residents using gas more than tripled, from 40 million to 150 million.
8. Data on per capita carbon emissions from fossil fuels are for 2002, from the United Nations Environment Programme database Global Environmental Outlook. Population data are for mid 2003, from the United Nations (2003).
9. The DRC data show no caps on smog and dust or carbon emissions. They show caps on smog only, but there is no disaggregation of smog and dust data for comparison.
10. China is positioned to leapfrog the world in nuclear power precisely because it entered the race late (*Newsweek* 2006).
11. This capability has been demonstrated for pulverized coal power plant, photovoltaic systems, and the like. Furthermore, the objective of the U.S. Department of Energy's IGCC program is to get unit costs down to US$1,000 per kilowatt in a smaller potential market in which construction costs are higher.
12. MARKAL, which stands for MARKet ALlocation, is a generic model, developed by the Energy Technology Systems Analysis Programme of the IEA, with an Italian university. The model is tailored by the input data to represent the evolution over 40 to 50 years of a specific energy system at the national, regional, state or province, or community level. The relevant Web site is http://www.etsap.org/markal/main.html.
13. This 100-megawatt wind farm project in Inner Mongolia is supported in part by the World Bank.

References

Alford, William P. 2002. "At a Crossroad: The Challenges of Implementing China's Environmental Law." *Harvard China Review* 3 (1): 10–13.

Communist Party of China Central Committee on Drafting the 11th Five-Year Plan. 2005. *The Proposals of the CPC Central Committee on Drafting the 11th Five-Year Plan.* Beijing: Communist Party of China.

DRC (Development Research Center). 2004. *Basic Conception of the National Energy Strategies.* Beijing: DRC.

Government of China. 2004. *The People's Republic of China Initial National Communication on Climate Change.* Beijing: Government of China.

Ho, Mun S., and Dale W. Jorgensen. 2003. *Air Pollution in China: Sectoral Allocation of Emissions and Health Damage.* Beijing: China Council for International Cooperation on Environment and Development.

IEA (International Energy Agency). 2004.

Murray, C. J. L., and A. D. Lopez. 1996. *Global Burden of Disease and Risk Factors.* Geneva: World Health Organization.

Newsweek. 2006. "China Leaps Ahead." February 6, pp. 34–38.

Qiao, Tianbi. 2005. "2005: China's Fresh Start." *China Today.* http://www.chinatoday.com.cn/English/e2005/e200512/p12m.htm.

SEPA (State Environmental Protection Agency). 2003. *State of the Environment Report.* Beijing: SEPA.

Stern, David I. 2003. "The Environmental Kuznets Curve." Rensselaer Polytechnic Institute, Troy, NY. http://www.ecoeco.org/publica/encyc_entries/Stern.pdf.

Stern, Nicholas. 2006. *The Economics of Climate Change.* London: Her Majesty's Treasury.

United Nations. 2003. *World Population 2002.* Geneva: United Nations.

World Bank. 1997. *China 2020: Clear Water, Blue Skies.* Washington, DC: World Bank.

———. 2001. *China: Air, Land, and Water—Environmental Priorities for a New Millennium.* Washington, DC: World Bank.

World Bank, China Ministry of Science and Technology, Deutsche Gesellschaft für Technische Zusammenarbeit, and Swiss State Secretariat for Economic Affairs. 2004. *Clean Development Mechanism in China: Taking a Proactive and Sustainable Approach.* 2nd edition. World Bank: Washington, DC.

Wu, Zongxin, Pat DeLaquil, Eric D. Larson, Wenying Chen, and Pengfei Gao. 2001. *Future Implications of China's Energy Technology Choices.* http://www.princeton.edu/~energy/publications/pdf/2001/Wu_01_Future_implications_of_China%27s%20energy-technology_choices.pdf.

CHAPTER 5

Securing Energy Supply

Key Messages

China's growing sense of energy insecurity is justifiable, but international experience demonstrates that security issues arising from geopolitical uncertainties, price volatility, and natural disasters are manageable. China's growing sense of insecurity comes as the nation quickly emerges from an era of energy autarky into one of increasing dependence on imported oil and gas, as threats to the integrity and the cost to the economy of energy supply arise, and as anxiety develops about the possible consequences of electricity deregulation.

Internationally, there were seven oil supply emergencies since 1950, and concerns about the security of oil imports therefore tend to dominate policy thinking. International gas supply, in contrast, has proven until recently more reliable, mainly because long-term export-import contracts entail interdependence. However, in January 2006, gas trade between the Russian Federation and Ukraine—which is essentially on a state-to-state basis—was interrupted for four days and then quickly and completely restored. Although this incident resulted from a complex of longstanding and well-known commercial and technical issues, it sparked European fears about the political security of Russian supply. As for electricity, lessons can be learned from blackouts in other industrial countries and the steps being taken to avoid them in future.

However, most experts recognize that supplies of oil and gas (and to a lesser extent other energy forms) will remain vulnerable to geopolitical uncertainties (aggravated since September 11, 2001); price volatility; and—as Hurricane Katrina recently reminded the world—natural disasters. Most industrial countries have devised and implemented mechanisms to enhance overall energy security, especially petroleum supplies. China needs to build on those experiences to develop both broad policies and specific measures to enhance the security of its energy supplies.

Energy supply uncertainties and risks can be mitigated and effectively addressed with a comprehensive national energy policy that stresses supply diversity, energy efficiency, renewable energy, and a more market-oriented oil and gas subsector, to attract needed investments and technologies. In terms of broad policies, China should more aggressively promote energy efficiency and indigenous resource alternatives to coal. Doing so would require further opening and better regulating its oil and gas markets, commercializing its national oil and gas companies, opening the subsector to international oil companies to attract investments and needed technologies, and developing its national sources—particularly clean and renewable energies. The country should clarify the security-enhancing roles of both international and national companies. Doing so would lead to the creation of a market-oriented, multisourced, robust national energy economy, which provides an important underpinning for security of supply.

The right mix of specific oil supply security measures should be selected according to China's needs from a suite of measures:

- Maintenance of spare domestic production capability
- Participation in international protection of oil flows through overseas chokepoints
- Accumulation of inventories above normal commercial levels
- Increased refinery flexibility
- Fuel-switching capacity in consuming sectors

- Allocation and possibly rationing systems to share scarce supplies equitably
- Close international cooperation with trading partners for whom secure oil supplies are essential for economic well-being and with energy exporters who have a similar interest in secure markets.

Recognizing the country's interdependence with global energy (and particularly oil and gas) and incorporating security of supply in the country's long-term energy strategy could be the first steps on the road toward a stable energy supply, one of the pillars of sustainable development for the sector and the overall economy during the coming decades.

China's Growing Sense of Insecurity

The growing sense of insecurity is global. Energy supplies can become insecure because of

- *Geological factors.* Factors such as depletion rates may cause physical limits on fuel supply, so that volume is no longer adequate to meet demand. An example is the Canadian east coast offshore gas fields that were brought to market in 2000 and that have produced much below expectations.
- *Political, military, and diplomatic factors.* Such factors can restrict energy supply, make it prohibitively expensive, or make it physically unreliable. Such factors affected most Iranian oil supplies during the labor strikes of 1978, the revolution of 1979, and the prolonged war with Iraq in the early 1980s.
- *Infrastructure bottlenecks.* Because of such bottlenecks, energy flows may not reach consumers. Energy supplies may no longer be competitive by international standards, as happened to most Western European coal industries in the latter part of the 20th century.

These generic causes of insecurity and their results, together with China's special concerns, are presented in figure 5.1 (see p. 90).

Figure 5.1 Energy Insecurity: Generic Causes, Effects, and China's Special Concerns

Main factors causing energy insecurity	Effects of energy insecurity	China's special concerns
Geological factors such as resource depletion causing physical limits on fuel supply	Inadequate supply	Rising oil imports that are outside China's political control
Political, military, diplomatic actions of suppliers	Unreliable delivery systems	Flaws in planning system that cause bottlenecks
Infrastructure bottlenecks	Uneconomic cost	Impact of power system unbundling on markets

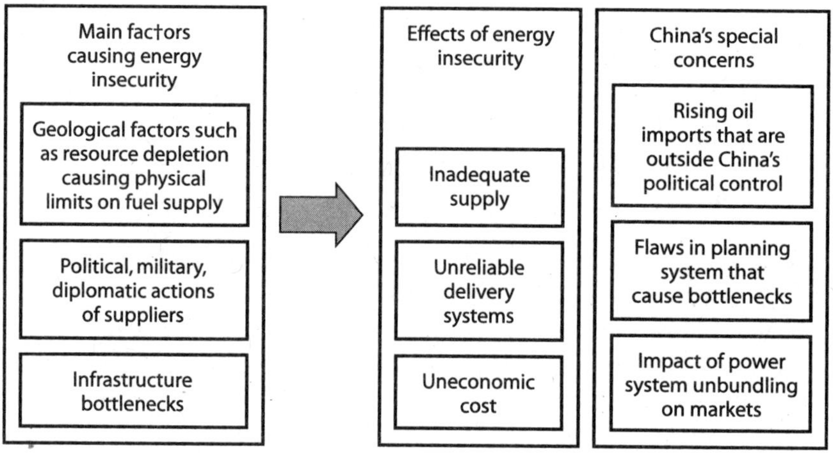

Source: Study team.

Key International Energy Insecurity Concerns and Their Causes

Since the middle of the last century, the two main global security concerns have been the possible disruption of oil imports—and to a lesser extent imports of gas or coal—and sudden price hikes, usually triggered by oil market events, which can threaten economic stability and competitiveness. For electric power systems, the "California syndrome"—or the association of power system failures with deregulation and competitive markets—has become a major concern. The September 11, 2001, attacks in the United States have raised concerns about the vulnerability of critical infrastructure such as oil and gas pipelines, power plants, oil refineries, and liquefied natural gas (LNG) terminals. The disruption of crude oil, oil products, and natural gas and electricity supply in the United States caused by Hurricane Katrina in September 2005 has focused attention on natural disasters and extreme weather events as a source of energy insecurity.

In addition to its military and political implications, energy insecurity can be costly in economic terms. Increased dependence on energy imports can raise foreign exchange requirements. The disruption of energy supply can stunt economic growth.[1] Furthermore, the need for security-enhancing measures to avoid such threats to economic growth can put a heavier burden on a country's budget.[2]

Oil imports. For the past half-century or more, international energy supply security concerns have focused on oil imports. To a large extent, this focus reflects the growing import dependence of many industrial countries, the concentration of imports from a relatively small number of suppliers, and the fact that on seven occasions those supplies have been significantly disrupted (see appendix G). Despite these disruptions, global oil markets have a good record over a long period of time in dealing effectively with both short- and long-term potential supply-demand imbalances, essentially through the price mechanism. Internationally competitive global trading countries with adequate foreign exchange have always been able to purchase all the oil they need at current market prices. An additional consideration in the first decade of the 21st century is the debate about peak global oil production, which may be bringing another element of uncertainty into policy discussions and long-term market behavior.

The risks surrounding petroleum imports are threefold:

1. Major exporting countries (for example, Saudi Arabia) or associations of such countries (for example, the Organization of Arab Petroleum Exporting Countries) could change their policies and constrain either total supply or the supply to particular markets.
2. Supply constraints could result from insufficient investments in exploration, field development, and transportation infrastructures. This scenario could occur either because national oil companies (NOCs) have inadequate investment funds or because governments restrict investments by international oil companies (IOCs) in some of the world's best areas for petroleum prospects (for example, the Islamic Republic of Iran, Kuwait, or Mexico).
3. *Force majeure* events could occur, such as civil disorder (strikes in Nigeria and the República Bolivariana de Venezuela); military actions (the Iran-Iraq War of 1981 to 1984); terrorist activity (so far, there has been only limited and local impact); and natural disasters (hurricanes in the Gulf of Mexico). These events could disrupt or interrupt oil or energy flows at regional or global levels. The perception is that since September 11, 2001, geopolitical uncertainties have been aggravated.

Gas imports. The rapidly developing international gas trade has been less subject to impairments than has oil trade—both of pipeline gas and of LNG. As major producer-exporters, Algeria, Canada, and Russia have been reliable suppliers for many decades. As consumer-importers, France, Germany, Italy, and (to a lesser extent) the United States have relied heavily on imported supplies to support strong domestic gas industries. Japan and the Republic of Korea have each confidently developed a large gas industry and related electricity generation capacity, almost wholly on the basis of imports. Turkey is following a similar path.

Where technical failures have occurred, such as accidental damage to LNG plants overseas, the industry has been able to quickly find alternative sources. International price disputes have arisen in the past—for example, between Canada and the United States, between France and Algeria, and between the Netherlands and other countries of the European Union (EU). However, since the 1980s, greater reliance on market-based pricing and better contracting practices have reduced the incidence of such disputes. Where international gas purchases involve long-term take-or-pay contracts, as is the case with most LNG trade, a valuable interdependence between sellers and buyers is created and contributes to security of supply.

The January 2006 dispute that resulted in a four-day disruption of supplies from Russia to Ukraine raised EU fears about the security of gas for Europe and led to a meeting of Group of Eight ministers in Moscow. Gas trade among the former Soviet countries is of a state-to-state nature. The causes of the dispute are complex, and some are longstanding. Prices are not market based. Other factors included Ukraine's payment arrears, the alleged diversion of gas destined for the European Union, and the rejection of an agreement for partial Russian ownership and for refurbishment of pipelines from Russia transiting Ukraine to Western Europe.[3]

Domestic energy sources. If domestic energy sources are not developed on an economic basis, they may not be secure, even though they are entirely under the control of the government. For example, the old, deep-mined coal industries that supported Western Europe's early industrial supremacy proved unable to meet the energy needs of a modern economy in terms of volume, quality, and price. Also in 1984/85, a one-year miners' strike in the United Kingdom caused the country's most serious peacetime energy crisis. More recently, strikes by petroleum transport workers and blockades to protest high prices for oil products have threatened the continuity of oil supplies to European consumers.

Technical failures may also impair the supply of domestic energy. Examples are accidents at a gas-processing plant in Australia and an LNG plant in Algeria and the failure of gas-producing fields in Canada (see appendix G).

Furthermore technical, financial, environmental, and regulatory issues are currently creating obstacles to needed gas development across North America, resulting in high prices and faltering consumption.[4]

A number of costly blackouts have occurred in the large, interconnected, but separately operated North American electricity systems. The most recent one, in August 2003, affected millions of customers and entailed significant economic losses. In the developing world, supply disruptions have occurred mainly because of insufficient generation capacity and weak transmission and distribution systems. These problems have been caused mainly by lack of investment but also by flaws in the planning, approval, or execution of projects. Countries affected include Argentina, Brazil, China, and India. Impairment of fuel supply for power generation can also disrupt electricity supply, as recently occurred in the southeast coastal regions of China.

Because domestic energy supplies are entirely within the nation's political control, as is the case for coal in China, they must in principle be regarded as more secure than imported supplies if they are developed on a sound economic basis and in harmony with global and regional markets. However, policy and regulatory steps need to be taken to avoid security risks. Such steps include subjecting these sources to market forces to ensure that they are internationally competitive, monitoring reserves and resources of fossil fuels related to long-term supply adequacy, addressing social issues that may affect supply, ensuring that investments are adequate to prevent infrastructure bottlenecks, and securing critical infrastructure against failures, however caused. A recent study sponsored by the World Bank, carried out in close coordination with the National Development and Reform Commission (NDRC), outlined a reform strategy in response to concerns about the future of China's coal development (see box 5.1 on pp. 94–95).

The Multidimensional Nature of Energy Security

Insecurity of energy supply relates to the nation's economy and society because consumption of commercial energy in its various forms is of enormous and pervasive importance. Security has a time dimension: harm can be caused by electricity disruptions that last a few days or by oil supply disruptions that endure for many months. It has geographic dimensions: disruptions may originate thousands of kilometers from the point of interrupted consumption or may arise very close to the consumption point. Supply can be disrupted by accident, sabotage, strikes or other social phenomena, terrorism, overseas wars and political actions,

> **Box 5.1**
>
> ## Bank Involvement in China's Coal Subsector
>
> **Coal: An Indispensable Source of Energy**
>
> China is the world's largest producer and consumer of coal and depends critically on coal to sustain its growing economy. With production and use approaching close to 2 billion tons per year, coal dominates the national energy market. China's coal subsector is not only the world's largest but also the most dangerous in terms of mine safety and the most polluting in terms of coal-fired boiler emissions. Coal is China's least costly source of energy. Despite official promotion of renewables and energy efficiency, coal demand will increase in the foreseeable future because of sustained economic growth and limited hydrocarbon reserves. Coal subsector reforms lag reforms in the rest of the economy. The development of small state-owned coal mines is unsustainable, and China is starting to look to the private sector to replace diminishing reserves and productive capacity. China is at a critical juncture for addressing a number of key challenges related to its large and growing coal industry.
>
> **Work to Date: An Emerging Reform Agenda**
>
> In response to concerns about the future of China's coal development, the World Bank sponsored a study funded by its Energy Sector Management Assistance Program to identify key issues and establish a constructive dialogue with the government of China through NDRC. This work included a review by a team of domestic and international experts of the institutional and regulatory framework governing the licensing and operation of coal sector enterprises in China. Coal industry practices were examined in two of the more important provinces, and two mine case studies were produced to demonstrate potential improvements. The findings were discussed at two workshops hosted by NDRC in 2003. These workshops were attended by a broad spectrum of stakeholders, and general consensus was reached on the formulation of a coal industry reform strategy to be sponsored by the World Bank. The strategy had these objectives:
>
> - Rationalize productive capacity and encourage private investment
> - Enhance occupational health and safety
> - Improve environmental protection.
>
> In July 2004, the World Bank issued a final report that consolidated this work (see World Bank 2004).

> **World Bank Support: A Potential Comparative Advantage**
>
> In many countries, the World Bank has supported efforts to introduce modern mining legislation, improve the investment climate, build institutional capacity to enforce laws, mitigate the social and environmental impacts of mining, and promote new investment and private sector development. Encouragement of positive social and economic impacts for local communities has been a growing component of World Bank interventions, one reinforced by the findings of the International Finance Corporation's Extractive Industries Review. The World Bank's comparative advantage and expertise in these areas could benefit China's new policy direction and coal subsector reforms. Preliminary discussions with NDRC suggest that these efforts could be directed toward improving China's coal industry by doing three things:
>
> - *Improving the management of coal resources.* The legal and regulatory framework could be modernized along the lines identified in World Bank (2004), and interministerial coordination could be improved.
> - *Reducing barriers to private investment.* Private sector access to geological and mining data could be improved, with mineral rights granted on an equitable and centralized basis, with security of tenure and increased operational autonomy.
> - *Enhancing institutional capacity.* Institutional capacity could be strengthened at all levels of government to deal with critical issues such as environmental protection and restoration of land for reuse, closure of uneconomical and damaging coal mines, and regional economic diversification to reduce dependence on coal mining and increase support for coal mine closures.
>
> *Source:* World Bank 2004.

and exceptional climatic events. Mechanisms to address insecurity are therefore multiple: improved training for energy system operators, technical repair capabilities, secure storage of some energy forms, and so on.

China's growing sense of insecurity stems from three factors:

1. Rising oil imports have placed nearly half of China's oil supply outside its political and military control, a situation dramatically different from that experienced only a decade ago when net import requirements were negligible.

2. Flaws in the planning system and transportation bottlenecks have threatened the integrity of energy (mainly power) supply, causing alternating cycles of oversupply and shortage.
3. The government's decision to deregulate its power industry has triggered concerns (the California syndrome) about the integrity of the power system.

Options for Securing Oil and Gas Supply

Long-term energy security requires vision and an integrated policy to support that vision. This section reviews measures widely used in the past, mainly by industrial countries, to secure oil and gas supplies. It includes a brief commentary and evaluation of what China is doing. China ought to consider these measures, evaluate them, and select the ones most appropriate to meet its needs.

Fostering the Development of National Oil and Gas Resources and Implementing the Renewable Energy Promotion Law of 2005

The development of national resources that can pass the test of international competitiveness and implementation of the Renewable Energy Promotion Law of 2005 (REPL) with more reliance on market forces is a foundational policy measure for securing long-term energy supply. Most countries with a heavy dependence on energy imports have focused primarily on policies to encourage domestic energy development. The most successful policies have targeted the development of resources that are economically and environmentally viable. Typically, new laws and amended policies have opened markets to attract capital and technology to energy businesses. They have provided incentives and created markets for the development of renewable energy resources and designed research and development and industrial strategies in new and renewable energy fields. According to the DRC's estimate, if China were to adopt this course of action as part of its general energy strategy, the share of domestic-source primary energy could still be as high as 80 percent in 2020, compared with about 90 percent in 2000.

China's oil and gas regime needs to be opened up. Comparative analyses show that the attractiveness of China's petroleum regimes are just average, compared with a worldwide rating of about 280 fiscal systems (World Bank and IESM 2000). Further changes are required to improve China's international competitiveness in several areas: the conditions for access to undeveloped resources, the fiscal terms for the development of such resources, and the conditions for the participation of domestic and international investors in exploration and development. Policy changes in these areas would result in more rapid development of domestic oil and gas and a consequent relative reduction in the nation's long-term import needs. Some examples of energy development incentives are given in box 5.2 (see p. 98).

China is encouraging a strategic transition of the oil and gas industry from the traditional areas in the northeast to Xinjiang and to offshore regions. Increased competition among the three major NOCs and improved access terms for IOCs, along with a general opening of energy markets, could help channel additional sources of capital and technology for the optimal development of China's oil and gas resources. However, long-term planning and supportive policies are clearly lacking in these areas.

China announced a very ambitious program for the development of renewable energy during the June 2004 Bonn Conference on Renewable Energy.[5] In addition, in late February 2005, China enacted the REPL and launched a competitive concession program to speed up the development of its sizable wind power resources. Then, at the Beijing International Renewable Energy Conference in November 2005, China announced an even stronger commitment to renewable energy, setting a target of doubling the current use of renewable energy to 15 percent of the country's energy balance by 2015. The World Bank and the Global Environment Facility are assisting China in this area through the China Renewable Energy Scale-Up Program (CRESP) announced in June 2005.[6]

Securing Imports through Diplomatic and Trade Relationships

Industrial countries have traditionally tried, though not always successfully, to attract foreign energy suppliers for the development of strong, secure trade relationships for energy (especially oil). France, Japan, and the United Kingdom seem to have been the most active in the Middle East, offering advanced technological and industrial assistance along with diplomatic support. Similarly, the United States tends to focus its energy trade relationships on the Middle East for oil imports and on Canada for gas and, increasingly, oil imports. U.S. companies play key technological assistance roles in some Middle East oil states.

The obvious focuses for China's development of these relationships are, for oil, with the Middle East—especially Saudi Arabia, which will continue to be a major oil supplier—and for gas, with Russia, the most important gas exporter. The importance of the emerging focus by China on Saudi Arabia for oil supply and on Russia for gas supply is clear from table 5.1 (see p. 99).

Box 5.2

Examples of Domestic Energy Development Incentives in Other Industrial Countries

Oil Sands

In Canada, development is encouraged by very low rates of taxation for oil sands compared with conventional oil.

Marginal Conventional Oil Fields

In the United Kingdom, the government grants tax relief, on a case-by-case basis to encourage investments that will help produce additional oil from partly depleted fields.

Natural Gas

In Canada, the federal government has subsidized the development of lateral natural gas pipelines and the expansion of distribution systems in order to reduce oil dependency, especially in remote areas.

Renewable Sources for Electricity Generation

In Texas, in the United States, a 1999 electricity restructuring law required 2,000 megawatts of nonhydrological renewable electric energy to be available by 2009. Already in February 2006, some 2,378 megawatts of wind energy alone were on line, according to the Texas Renewable Energy Industries Association, and Texas has overtaken California as the leading wind energy state of the United States (see http://www.treia.org/general_info.htm).

Wind Power

Following the energy crisis of 1973 to 1974, the United States became the world's largest producer of wind energy. However, since the early 1990s, Denmark and Germany have taken the lead in this area. Both countries have market-based energy policies with a high degree of freedom to invest in, sell, and buy energy without government interference. Tax incentives encourage both investment in wind power and consumption of electricity derived from it. For example, Germany has an eco-tax on all forms of traditional energy and uses the revenues to foster renewables. Currently, Germany has some 15,000 megawatts of wind power generation capacity installed. Denmark provides income tax incentives for investors in wind energy and is second in the world in terms of installed capacity from that source.

Source: Authors.

Table 5.1 Oil and Gas Reserves in the Middle East and Russian Federation

Country	Oil			Gas		
	Reserves (billion barrels)	Share of world reserves (%)	Reserves/ production (years)	Reserves (trillion cubic feet)	Share of the world reserves (%)	Reserves/ production (years)
Saudi Arabia	262	25.0	86	n.a.	n.a.	n.a.
Russian Fed.	n.a.	n.a.	n.a.	1,660	27.0	81
China	24	2.3	19	64	1.0	17

Source: Study team based on data from BP Global 2005.

Note: n.a. = not applicable.

China appears to be making vigorous efforts to use its diplomatic influence and trading strengths to improve relations with oil-exporting countries and thus to secure access to investment opportunities and trade flows. This effort is characteristic of the normal global competition between buyers of raw materials, in which China is now taking part. This sort of competition has existed for decades between other industrial countries such as those of the EU and Japan. It can take place in ways that do not conflict with the ongoing consultation and cooperation that is necessary between oil importers in organizations such as the International Energy Agency (IEA) and the Asia Pacific Economic Cooperation (APEC).

Building Relationships with the IOCs

Energy is the world's biggest business and the domain of its largest corporations. Among the IOCs, ExxonMobil is the world's largest private corporation. Among the NOCs, Saudi Aramco is the world's largest oil company, with reserves some 20 times greater than those of ExxonMobil. Close relations with these and similar companies can help an importing country access international oil supplies. Indeed, many industrial countries have relied very successfully on the IOCs to secure all their import needs. IOCs can have a useful (import) role in China, and their involvement could be considered along with that of Chinese NOCs to enhance security of supply. Relationships with the IOCs would also provide access to state-of-the-art technology and management practices across the whole range of oil and gas industry activities.

The IOCs are unequalled in their relationships with producing countries; their ownership of equity crude oil; and their control of transportation infrastructure (shipping, pipelines), which gives them great global flexibility in their oil supply operations. They are involved in all segments of the oil and gas business, from exploration and production to retail distribution. They are managerially strong and are experienced in risk-hedging techniques, both areas in which the Chinese NOCs may be relatively weak. However, they currently control only 10 percent of the world's oil and gas reserves. Of the top 20 owners of oil and gas reserves in 2004, 14 were NOCs. State monopolies are the 10 largest owners of such reserves. ExxonMobil and the Shell Group rank 12th and 13th. BP and Chevron-Texaco are 16th and 19th.

Defining a Role for the NOCs

NOCs now are out of fashion in most industrial energy-consuming countries that dabbled with the concept in the past half-century. For example, the U.K. government long ago gave up its share in BP. The later British National Oil Company, which the government created to explore and develop in the North Sea, has been privatized. Petro-Canada has been almost completely privatized, the government retaining only about 14 percent of the share capital. In France, the former state company Elf is now part of the large, private TotalFinaElf group. The Spanish state company Repsol was successfully privatized, and then it absorbed the privatized Argentine oil monopoly Yacimientos Petrolíferos Fiscales and, as a result, has become a significant international player, particularly in Latin America. Japanese policy makers appear to be disillusioned with their investments in the Japan National Oil Company. In fact, among the IEA members, only the Italian state company Eni appears to be a significant NOC player.

The Chinese NOCs, with government support and encouragement, have pursued oil and gas production and export investments in many countries. However, a rapid examination of these activities suggests that these investments are made ad hoc. Some of the bilateral deals seem to have been initiated and conducted without proper analysis and due diligence, perhaps guided by an overreaction to emerging oil security concerns. Moreover, as a consequence of their listings on international markets, the Chinese NOCs appear to be facing the dilemma of either pursuing strictly commercial investment strategies or taking into consideration broader objectives of energy security for China. The role of the NOCs in supporting China's energy security objectives seems not yet to have been defined.

The NOCs can contribute significantly to China's oil and energy security by providing access to adequate and secure oil and gas supplies. However, the partially listed companies are increasingly coming under investor scrutiny. A perception that they give precedence to China's national interest over their shareholders' interest could hurt their market standings. Therefore, the government should define and publicize the role of the NOCs in enhancing the country's energy security. If they are required to take on operations that are suboptimal from the standpoint of maximizing returns to shareholders, the NOCs should receive adequate and transparent compensation. NOCs also are beginning to face scrutiny over business and environmental practices, including investments in countries at high political risk. This scrutiny will intensify with their increasing international visibility and should be taken into account in devising their global investment strategies and business behavior. One aspect of current NOC activity directed at improving supply security—the repatriation of equity oil—is highlighted in box 5.3 (see p. 102).

Clarifying the Security Roles of Foreign Oil Companies

It is timely for the government to similarly clarify the roles that it expects foreign oil companies, both private and state owned, to play in the Chinese market. In particular, there should be clear definition of conditions for market entry and exit and for partnership with Chinese NOCs and other local operators in such areas as importing and storage (including strategic storage), exploration and development, refining and processing, and distribution and marketing. All policy and regulatory conditions should be transparent and be applied in a nondiscriminatory manner. Opening markets to enable IOCs and other foreign investors to play a greater role in oil supply and domestic resource development could be a positive signal to the international community, making it easier for Chinese NOCs to engage in similar activities abroad. China's NOCs should at the same time consider strategies for interacting profitably with and learning from foreign oil companies, both NOCs and IOCs, as well as strategies for better integration in global and regional oil markets. All major players recognize and seem to welcome China's increasing role in the oil market and, to a lesser extent, in gas markets. The investments of Chinese NOCs in the development and marketing of new reserves are perceived as a stabilizing factor in future oil and gas markets. More transparency would certainly improve the business orientation of Chinese NOCs and ease their entry into the global market, where they could be world-class players.

Box 5.3

Repatriating Equity Oil: Is There a Better Way to Provide Security?

One aspect of the current activities of China's NOCs calls for a specific comment. It appears to be their policy—and possibly that of the government—that their equity (or "owned") crude oil that is found and developed abroad be repatriated to China for refining, almost regardless of its location or quality. The outstanding case is oil production by the China National Petroleum Corporation in Sudan: it is exported to China for refining rather than to the closer south European market.

Argument for Repatriation

This flow of oil is under Chinese control from the point of loading in Sudan to the distribution of the final refined product. Hence, the oil is more secure than any foreign barrels purchased on the international market.

Argument against Repatriation

There may be an added cost to China's NOCs—and therefore to the Chinese economy—resulting from the longer distances involved and the need to dedicate refineries or refinery units to processing so-called difficult grades of crude oil. It is the practice of IOCs to maximize the value of their oil production by selling at the highest available price and to minimize the costs of their refinery feed streams by purchasing at the lowest available price, regardless of the nationality or ownership of the seller or buyer in each case.

Alternatives

One possibility is to examine whether it might make more sense to sell Sudanese oil to southern Europe and buy other barrels for China's needs. Another is to consider exchanging the Sudanese oil for oil from a closer source, making the supply of the Sudanese barrels to the exchange partner contingent on receiving the exchange barrels in China. That comment is made while recognizing that there may be technical or trade obstacles, now or in the future, to delivering Sudanese oil to certain markets.

Source: Authors.

Establishing Policies to Enhance Short-Term Oil Supply Security

International experience clearly indicates that the accumulation of oil stocks above normal commercial levels is an important and worthwhile component of an integrated approach to improving the security of China's oil supply. There are several key issues to address in any stockpiling program. First, where, how, by whom, in what amount, and in what form should the stocks be held? Second, how should the costs of stockpiling be met? Third, and perhaps most important, under what circumstances will the government release the oil stocks? In most industrial countries, these issues are a matter of public policy. Appendix H, a summary of a paper prepared by the World Bank in January 2002, discusses these issues.

The contribution of normal commercial stocks. The oil industry holds commercial oil stocks to deal with fluctuations in demand and to protect against technical failures and operational delays that could affect the continuity of supply for customers. Typically, oil companies attach much importance to keeping their customers supplied. They are also used to working cooperatively to assist each other in the event of technical failures. However, like many other manufacturing and distribution businesses, oil companies around the world have been progressively reducing their inventories to control costs. They do not plan their inventories taking into account supply emergencies caused by geopolitical events. To help address such contingencies, either governments themselves hold reserves in strategic storage or they mandate that the oil companies maintain minimum stock requirements. Some governments adopt both measures.

Strategic oil storage. In the late 1990s, the government of China announced its intention to keep crude oil reserves equivalent to 60 days worth of net imports and 30 days worth of oil products consumption by 2020. China National Petroleum Corporation (CNPC) has begun work on strategic oil storage. Industry analysts have assumed that a portion of the large increases in China's oil imports in recent years has been for strategic storage purposes.[7] There is no question that accumulating stocks over and above minimum commercial requirements is the most important and valuable step that China can take to avoid negative economic and social impacts of short-term oil supply disruptions. Several oil-importing countries maintain strategic reserves as one element of civilian oil supply security policies; examples include Germany, Japan, Switzerland, and the United States. Reportedly, India has approved an oil supply security plan modeled on the U.S. Strategic Petroleum Reserve.

Current high steel prices[8] have made storage in steel containers—tanks or barges—more expensive. The cost of this storage may be a charge on the general budget as in the United States, or the cost could be recovered by a security tax on oil consumption. Strategic petroleum reserves are attractive to policy makers because they are known quantities that are available at any one time in given locations under their control. Moreover, policy makers control the rate at which withdrawals from these stocks can take place in order to achieve policy goals. For example, the U.S. Congress approved the U.S. Strategic Petroleum Reserve in 1975, in the aftermath of the oil crisis of 1973 to 1974; it currently covers about 59 days worth of imports.[9] Japan's strategic petroleum reserve is a minimum of 50 million cubic meters, equivalent to about 55 days of consumption at 2006 rates. Table 5.2 ranks strategic oil storage and other key measures for short-term oil supply security in terms of their difficulty and relative cost.

A strategic petroleum reserve clearly gives the government the greatest degree of control over the location, levels, quality, physical security, and use of an emergency oil inventory. However, developing a strategic petroleum reserve is an expensive proposition, especially with crude oil prices in the second half of 2006 fluctuating in the US$60 to US$75 per barrel range. A policy decision would be required as to how the costs of a reserve could be met. For example, they could be paid out of general government revenues, as in the United States, or they could be recovered from oil users by a levy on consumption.

Table 5.2 Assessment of Relative Difficulty, Cost, and Degree of Control of Key Measures for Short-Term Oil Supply Security

Key measures for short-term oil supply security	Relative level of difficulty	Relative cost to government	Degree of government control
Strategic petroleum reserve	Low	High	High
Mandated minimum commercial stocks	Low	Low	Medium
Spare domestic oil production capacity	Medium	High	Medium

Source: Authors.

Mandated minimum commercial stocks. Many countries have laws and regulations that require their oil companies to maintain stocks at certain minimum levels relative to consumption. For example, consistent with the EU directive on stocks, most EU member countries require their industries to maintain minimum stocks. These stocks correspond to 90 days worth of the average internal consumption in the preceding year of the three main categories of oil products. Countries that are in a net export position, such as Denmark and the United Kingdom, do not have the same obligation. The United States does not mandate minimum commercial stocks, but the U.S. Strategic Petroleum Reserve plus commercial stocks are currently equivalent to about 118 days worth of imports. Japan requires companies to hold commercial stocks equivalent to 70 days worth of consumption, over and above the volumes held in Japan's reserve.

Generally, oil companies are authorized to recover from consumers the costs of mandated minimum commercial stocks. If governments apply stockholding requirements equally to all companies and if the companies experience roughly equal cost increases as a result, they are likely, in a market economy, to be able to recover these costs through the selling prices of oil products. The attraction to policy makers of mandated minimum commercial stocks is that there is no budgetary cost, and yet, at least in theory, these stocks can be drawn down only at the direction of the government.

Stocks owned by oil exporters and held in oil-importing countries. Exporters also have a strong interest in avoiding supply disruption in their important markets. Saudi Arabia reportedly maintains a huge storage capacity near Europe and in the United States, in order to ensure timely and cost-efficient delivery of supplies to those markets. There do not appear to be publicly available data about the current quantities, qualities, and locations of the stocks. Also, Saudi Arabia reportedly has sought storage capacity in the Asia-Pacific region but has been unsuccessful because so far no government in the region has allowed Saudi Arabia to set up facilities. There may therefore be a mutually beneficial opportunity for a Chinese initiative in this area.

Maintenance of spare domestic oil production capability. This option is costly because, to be effective, it requires the creation of redundancies at many stages of the domestic oil supply chain—the capacity of wells, processing plants, pipelines, and possibly refineries processing crude oil. A review of international experience shows that many countries have not considered it a worthwhile option because of the substantial investment and maintenance necessary. It is mentioned here only because at some stages in the development of the upstream industry such linked spare capacity may exist that could be called on in an emergency.

Two cases illustrate this measure. The first case occurred twice in Texas, United States, during the Suez crisis of 1956 to 1957 and the Suez crisis of 1967. The regulator, the Texas Railroad Commission, gave approval for additional volumes to be produced in the domestic market, which reduced import needs and allowed the diversion of available foreign oil, such as that from the República Bolivariana de Venezuela, to Europe. The second case is Canada's response to the emergency circumstances of 1973 and 1974. Canada's developed wellhead and crude oil pipeline capacities were greater than those of the plants processing associated gas. Therefore, the regulator was prepared to consider permitting the flaring of associated gas in order to bring this spare wellhead crude oil capacity into production.

It is unlikely that there is significant unused oil production capacity in China's domestic oil industry under any normal circumstances. However, it is important to evaluate what, if any, potential may exist in emergency conditions and to keep track of the development of that potential. An assessment should be made as to whether China's production companies could, in an emergency situation, increase production from existing oil wells, where surface facilities such as gas processing or pipelines do not match the developed wellhead capacity. Increasing production might involve, for example, flaring associated gas or transporting extra oil production by what would normally be regarded as noneconomic methods, such as long-distance transport by rail or truck. The assessment should be updated from time to time.

Provision of military protection for crude oil traffic in the chokepoints of international tanker transportation. Disruption of tanker traffic in the narrow Hormuz, Malacca, and Sunda straits could affect the reliability or cost of China's oil supply. An interruption of commercial access to the Malacca and Sunda straits would significantly increase the amount and cost of shipping capacity needed to supply Middle East oil to points east of those straits.[10] If this situation were to occur when the international tanker markets were already stressed, it could be very difficult for the industry to compensate for the shortfall in crude oil supply.

Major nations having an interest in the security of global oil supply—and China is certainly one of them—could cooperate in devising measures to ensure that illegal activities in these narrow waters do not disrupt petroleum supplies.[11] Information on these matters is generally not public, and its evaluation is beyond the scope of this report.

Building Flexibility into the Oil-Refining Industry

Oil refineries are the critical interface between crude oil supply and the oil products needs of consumers. Typically, refineries are built and operated to handle a predetermined range of crude oil types. That range may be narrow, especially if the refineries are associated with petrochemicals and manufacture of lubricants. Generally, capital costs are lower for refineries that are designed to handle a narrow range of crude oils (in terms of, for example, sulfur content and specific gravity) than for plants that can process a wide variety of feedstocks, including high-sulfur ("sour") heavy crude oils. However, it happens in times of crisis that unusual grades of crude oil may be available when traditional sources are not. In these circumstances, refinery flexibility can be a valuable asset from the standpoint of ensuring the continuity and security of oil products supply.

To the extent that the Organization of Petroleum Exporting Countries (OPEC) has spare production capacity, it is often in the form of heavy and relatively sour crude oils. For example, in normal circumstances, Saudi Arabia maximizes the production of its Arab Light grade of oil. In emergencies, when larger volumes are necessary, it usually maximizes production of the Arab Heavy grade. Refiners who are able to handle this grade of crude and even less desirable grades of oil clearly have an advantage over refiners with less flexible equipment.

A lack of adequate refining capacity to handle heavier and more sulfurous crude oils appears to be an ongoing issue in China's downstream subsector. Several refineries are being upgraded to handle heavier and sourer grades of oil. These programs should be accelerated, because they will provide access to a wider array of supply sources. They can also result in much improved profitability for refineries that do undertake the investments needed to process heavy sour crude oils. The rise in international crude oil prices in the past five years has been particularly pronounced for grades of oil with high specific gravity and low sulfur. The market price differentials of these grades compared with those of higher specific gravity and more sour crude oils have greatly increased. As a result, the profit margins available to refiners of the more abundant, poorer-quality crude oils have been very large.

One major project undertaken by China in this field is the joint venture between Saudi Aramco, ExxonMobil, and Sinopec to expand and rehabilitate the Quanzhou oil refinery in Fujian province, to be able to process heavier sour Saudi crude oil, among other grades. This initiative should be followed through and extended, and policy makers could consider linking further incentives and encouragement to build even greater flexibility into the refining industry's processing capability. Greater technical flexibility would enable refineries to seek out poorer-quality crude oils that may be more readily available in an emergency to meet the market's needs.

Maintaining or Developing Fuel-Switching Capabilities in Large Consuming Industries

Fuel-switching capabilities often exist in large energy-consuming industrial plants in countries that have deregulated energy commodity markets. For example, this capability often exists in thermal generating stations in areas where different fuels—coal, fuel oil, and natural gas—are available. Many owners of such plants have found it worthwhile to make the additional investment in order to be able to play the market for different fuels. Fuel-switching capability is usually considered for commercial reasons, but it could be encouraged and used to secure supply in emergency cases.

One geographic area where fuel switching for power generation has been important is in the New England area of the United States, where switching is used for electric power generation. Where such equipment exists, it can usefully add to the security of oil supply by enabling the release of oil or gas to other consumers who do not have such capability. However, it may be costly to build some types of fuel-switching capability into a plant, especially because fuel suppliers may have to make related investments. For example, it may require investments in gas pipelines that remain underused at times. The economic case for investment in such capacity requires careful consideration if the switching capability is specifically for security reasons.

Such options have not been considered or even explored in China, probably because of the strict controls on prices and fuel supply and because project approvals depend on initial capital cost, without consideration of risk mitigation and operational flexibility measures. However, if power subsector reform progresses as envisaged and if consumers have direct access to energy providers, international experience shows that they will consider such options to secure access to different segments of the energy market, thus increasing their bargaining power with energy suppliers.

Setting Up Emergency Allocation Schemes

It is prudent for governments to have priority allocation schemes in times of scarce supply. These schemes are simpler and cheaper to operate than rationing programs. Their purpose is to ensure that vital services—police and fire services, ambulances, and hospitals and clinics—have adequate oil supplies in the event of modest shortfalls. A second level of priority might be industrial and commercial transportation needs. At the lowest level might be fuel for discretionary travel requirements. Canada developed an allocation system in the 1970s and revised it in the early 1980s under the authority of its Energy Supplies Allocation Board. Although Canada has since mothballed the system, it could be readied for use at relatively short notice.

No information is available on what China may be doing about emergency allocation programs. The international consensus seems to be that an emergency allocation system is a desirable element of national measures to deal with temporary oil or gas shortages. Allocation systems can minimize economic losses and disruption to vital social and emergency networks and are relatively easy and inexpensive to create.

Establishing Rationing Systems for Oil Products

A rationing system, which would require that consumers present both cash and rights in order to acquire oil products in an emergency, is a complex and controversial project in modern industrial societies. No doubt it could be managed more easily now, using information technology, than it could be in the era of ration coupons. Nevertheless, it may be too complex a project to embark on after an emergency has already occurred.

In developing a comprehensive energy policy framework, the government needs to decide whether it is worth designing a rationing system for oil products to be held in reserve for emergency use. An alternate, and probably more suitable, approach is to rely on the price mechanism for rationing, combined with a limited allocation program that ensures priority for the fuel needs of emergency services.

Engaging in International Cooperation for Short-Term Oil Security

China is now recognized as a major player in world oil markets, and its influence is increasing in natural gas markets. Most industrial countries and major industry players in the market see the need to engage, rather than antagonize, China. This attitude is reflected in the way in which the IEA, the world's leading energy consumer group, and Saudi Arabia, the major producer in OPEC, are both engaging the government of China to address global issues related to energy production and consumption. Political will on all sides should focus on mutually beneficial cooperation that can help bring stability to oil and gas markets. The following paragraphs briefly address areas of actual and potential global, regional, and bilateral cooperation.

Global multilateralism: The International Energy Agency. The oil market is a global one. Any major disruption of supplies that is significantly greater than the worldwide industry's spare production capacity, that lasts for a substantial period of time, and that exceeds the industry's immediate capacity to draw down stocks will therefore have global impacts on oil volumes and prices. In today's interdependent world, no trading country is likely to avoid the risk of economic harm, no matter how well prepared it may be in terms of strategic stocks. Also, countermeasures to deal with the causes and effects of oil disruption, such as price spikes, are much more effective if taken collectively. The Organisation for Economic Co-operation and Development has recognized these two factors, which underlie the creation, in November 1974, of the IEA[12] out of a desire "to promote secure oil supplies on reasonable and equitable terms."[13]

The member countries of the IEA commit to hold emergency oil stocks equivalent to 90 days worth of net oil imports. They also agree to take effective cooperative measures to meet any oil supply emergency (see box 5.4). At the same time, over the long term, members strive to reduce their vulnerability to an oil supply disruption. The means to attain this objective include increased energy efficiency; conservation; and development of coal, natural gas, nuclear power, and renewable energy sources, with a strong emphasis on technology.[14]

Formal consultation of a general nature between China and the IEA began in 1994. A memorandum of policy understandings was signed in 1996. The IEA continues to pursue technical contacts and dialogue with those responsible for China's energy policy. The IEA and China have maintained a liaison on security matters and emergency response, particularly since 2002.

> **Box 5.4**
>
> ## The IEA's Approach to Short-Term Oil Emergencies
>
> The IEA maintains two approaches to dealing with short-term international oil emergencies.
>
> The first, created in 1974, is the Emergency Oil Sharing System (EOSS). The EOSS is based on predefined triggers that activate the system and uses prescribed arithmetic approaches to determine how scarce supplies of imported oil should be allocated among members of the IEA. There have been several paper exercises of the EOSS, and it has been triggered once, for a short time and on a very small scale. Currently, the system is regarded as too mechanistic and inflexible for use in the event of a major international supply crisis.
>
> The second approach, the Coordinated Emergency Response Measures (CERM), was developed in 1984 to take account of lessons learned during the Iranian revolution and the Iran-Iraq War. The CERM approach includes demand restraint, surge production, fuel switching, and equitable allocation of scarce supplies among member countries. These measures now take priority over the more narrowly based EOSS provided for in the International Energy Program agreement. When the CERM approach was reviewed in 1994, IEA ministers expressed the view that coordinated stock draws can be a rapid and effective means of restoring interrupted supply, particularly in the early stages of a disruption, but that effective decisions on the stock draw cannot be made in advance of the disruption.
>
> The IEA's Standing Group on Emergency Questions provides an international forum for discussing emergency issues, for planning to deal with them, and for exchanging information. This Joint Oil Data Initiative—which involves governments, IOCs, and NOCs from IEA and non-IEA countries, OPEC, other producers, and consumers and regional associations—provides information exchange on a broader scale. The point is that everyone becomes more secure.
>
> *Source*: Based on information from the IEA Information Centre, http://www.iea.org/Textbase/subjectqueries/keyresult.asp?KEYWORD_ID=4103.

Consultation and cooperation with the other industrial countries—bilaterally and, particularly, in the IEA context—is clearly important and valuable for all parties. China does not need to be a member of IEA for this activity to be effective in helping meet the country's supply security objectives. The same is likely true even in the event of a global supply emergency, when coordinated actions with international organizations can be achieved without formal membership. However, regional markets could face localized disruption or security issues that require specific measures. East Asia—and more generally Asia—contributes the lion's share in the growth of world energy demand, particularly in the oil and gas markets. The common interests of the East Asian countries require intensified regional cooperation to reduce the cost of energy security and take advantage of their proximity to regions well endowed with supplies.

The industrial countries, individually and through the IEA, attach high value to international energy and oil cooperation for two principal reasons. First, they generally subscribe to the view that markets can work effectively for energy and for oil as for other commodities and that barriers to efficient energy trade should be removed to enable the optimal functioning of markets. Second, they have had favorable experience with such cooperation; for example, in 1990 and 1991 during the Iraq-Kuwait War. The IOCs understandably take a similar view. The Shell Group has recently tried to quantify the outcomes, in terms of global economic growth, of market-oriented versus alternative approaches to dealing with energy security, while addressing social justice and efficiency objectives (these are what Shell calls a "trilemma"). Its conclusion is that global economic growth would be highest where energy security is achieved by heightened investment, consumer countries strike win-win deals with producers, and energy prices include the cost of environmental effects. This work is summarized in box 5.5.

Regional multilateralism: Asia Pacific Economic Cooperation. APEC, a younger, less structured organization than IEA, is a group of circum-Pacific countries, which includes such energy heavyweights as Canada, China, Indonesia, Japan, Mexico, and the United States.[15] Energy issues, including security of supply, have attracted much attention, from the ministerial level down to subcommittees. So far, specific joint actions to improve security have not been agreed on, possibly because key members—Australia, Canada, Japan, and the United States, for example—see global multilateralism as the most effective means to address supply security. They are therefore already heavily involved in the IEA and might see APEC activity as possibly diluting the IEA's key role here.

Regional multilateralism: Association of Southeast Asian Nations. China is not a member of the Association of Southeast Asian Nations (ASEAN) but has from time to time taken part in its activities.[16] For

> **Box 5.5**
>
> ## The Shell Group's Latest Energy Security Scenarios
>
> Scenarios are "stories of equally plausible futures" (Schwartz 1996, p. xiii). They help governments and corporations make strategic decisions that will be sound for all plausible futures. The Shell group of companies developed scenario planning in the 1970s (Schwarz 1996). In June 2005, Shell launched new global scenarios that look forward to 2025. The search for energy security remains a key consideration, in Shell's view, which is summarized in the following points (Shell Group 2005).
>
> - Security covers physical supplies threatened by international insecurity and depletion of supply sources.
> - Insecurity can arise from lack of investment in existing energy sources, new energy sources, and infrastructure.
> - Approaches to addressing energy security are affected by different attitudes to globalization:
>   - —In *low-trust globalization*, security is achieved by proactive policies that diversify supply and reduce vulnerability to external shocks, but OPEC supports high prices and restricts investment (global growth: 3.1 percent).
>   - —In *open doors*, security is achieved by increased investment, win-win deals between consumers and producers, and energy prices that internalize environmental effects (global growth: 3.8 percent).
>   - —In *flags* (nationalism), security becomes a routine part of diplomatic and military relations; consumers trade secure markets for secure supplies (no growth estimate).
>
> *Source:* Schwartz 1996, p. xiii.

example, China, together with Japan and the Republic of Korea, which are also nonmembers, attended a joint ASEAN energy ministers' meeting in June 2004. ASEAN ministers welcomed China's initiative to establish a national oil stockpile program, looked forward to the technical assistance of Japan and the Republic of Korea on stockpiling matters, pledged concerted efforts to address regional oil market issues, and promised to address issues in natural gas development. Ministers also said they would pursue dialogues and partnerships outside the region, particularly with Middle Eastern oil-producing countries. These issues are all areas of interest and importance to China.

Regional multilateralism: Northeast Asia Energy Cooperation. There may be a valuable opportunity for regional multilateralism of a different sort in northeast Asia among countries such as China, Japan, the Democratic People's Republic of Korea, the Republic of Korea, Mongolia, and Russia. Such cooperation could, for example, harness the markets, technology, and capabilities for capital accumulation of China, Japan, and the Republic of Korea to the development of Russia's Siberian and Pacific energy sources. It would involve technology transfer, investment in energy development, transit arrangements with Mongolia and eventually the Democratic People's Republic of Korea, transportation infrastructure, and consequent trade flows. This security-enhancing cooperation is, of course, different in kind than the policy measures and joint actions in emergency circumstances that have been developed by the IEA. A successful international legal framework for investment and trade in energy, resulting in enhanced confidence, has been developed under the Energy Charter Treaty.[17] The treaty has been ratified by Japan, Mongolia is a full contracting party, and China and the Republic of Korea have observer status. One upshot of the January 2006 Russia-Ukraine gas dispute, which entailed transit pipeline issues, was a request that Russia should ratify the treaty, something that the Russian government has promised to examine.

Bilateral international relations. It is clearly important for China to continue developing and enhancing strong bilateral relations with energy-exporting nations. Bilateral relations are important not only in the Middle East, from which China presently obtains about half its imports, but also in Russia (Siberia), the Central Asian countries (Kazakhstan and Turkmenistan), the Caucasus (Azerbaijan), and Africa (CNPC is the operator and major owner of the largest oil-producing business in Sudan). To judge from the level and frequency of contacts, Latin American countries such as Brazil, Mexico, and the República Bolivariana de Venezuela are all becoming important partners in China's international energy policy. Diplomatic initiatives are often followed by investments by China's NOCs. In this context, CNPC now operates in 32 countries. Not surprisingly, Chinese interests in such areas as Russia, Central Asia, Africa, and the Middle East come up against powerful economic competition.

Bilateral relations are also important with other Asian importing countries, such as India and Pakistan. A reasonable goal for this relationship is not simply to make the supply of energy more secure but to manage, with other Asian giants, the geopolitical implications of China's large and growing energy needs.

Options for Securing Electricity Supply

Despite the tremendous growth of installed electricity capacity in China and the expansion of the transmission system since the early 1980s, the reliability of electricity services in China remains low compared with systems in the industrial countries and even some developing countries. Though difficult to quantify, the economic losses that resulted from power shortages in the 1980s and from 2003 to 2006 are significant, especially in the high-growth regions of the country. For example, it is estimated that economic losses caused by power shortages in Zhejiang amounted to about Y 100 billion in 2004.

In March 2002, the State Council issued a decree separating generation plants from transmission and distribution systems. The result, in the absence of well-functioning markets, is a greater disconnect between the development of generation capacity and the development of transmission networks. The events in the United States of September 11, 2001, brought policy makers' attention to security issues related to intentional damages to generation and transmission installations, beyond the vandalism and pilfering of equipment that often takes place in developing countries.

Securing electricity supply requires increasing the reliability of the electricity industry in China, through more effective and coordinated planning beyond the requirements of the five-year plan for generation, transmission, and distribution. In addition, China needs to improve operation and maintenance procedures as well as develop measures to limit intentional damage to equipment and ensure rapid restoration in case of deliberate attempts to destroy them.[18]

Improving Planning and Maintenance Procedures

Despite some highly visible blackouts, the reliability of electricity services in industrial countries improved significantly during the past decades. Reliability improvements stemmed mainly from technological progress, better interconnection of the systems, increasingly sophisticated forecasting and system planning methodologies, and greater focus on preventive maintenance. However, in recent years, utilities in industrial countries have begun to face serious challenges from customers, such as information technology firms and services providers, whose equipment is particularly sensitive to the quality and continuity of electricity services. These challenges clearly indicate the need to tailor reliability criteria to the specific conditions of the system and the needs of clients.

Improving the reliability of electricity services in China requires improved planning procedures at the generation, transmission, and distribution levels. It also requires increased reliance on preventive maintenance and more flexibility in the project approval process, to allow utilities to better respond to clients' needs for more reliable electricity supply. In China, electricity generation planning suffers from a greater focus on supply than on demand, ad hoc and project-based approval procedures, and a lack of economically and technically justified criteria. Project approval rarely is based on comprehensive, long-term system development studies. Forecasts focus on energy (kilowatt-hours) rather than on loads (kilowatts) and are usually limited to the time horizons of the five-year plans, with no possibility of adjustment during the plan period. These forecasts are based on normative electricity consumption per unit of product for the main industrial sectors, which still account for more than 70 percent of total consumption.

The planning process does not take into account structural changes of load curves, and the result often is less than optimal investment in peaking capacity. Engineering criteria rarely result from cost-benefit analyses and are not adapted to the specific conditions of the systems. As a result, power systems experience successive boom and bust periods and chronic power shortages. The recent adoption of better planning processes at the utility level has not brought about the expected improvements in reliability. Most provinces have experienced power shortages because of inadequate and short-sighted forecasting, combined with unchanged project approval procedures.

Reliability issues in China are compounded by underdeveloped or inadequately designed transmission systems, mainly because of weaknesses in the planning procedures or the existence of price controls. In China, transmission and distribution systems are developed on an ad hoc basis because they are funded mainly through connection fees. Utilities tend to minimize funding from their own resources or borrowing because strict price controls do not allow them to pass full costs to end users.

The separation of generation from transmission and distribution in China has further deteriorated the power system's planning environment because of the lack of cooperation between generators and network companies. If not addressed through proper regulation, the situation could deteriorate even further. Recent experience in the United States indicates that with the increased deregulation during the 1990s, the development of transmission capacity lagged the growth of generation capacity. One reason for this disconnect was the reluctance of the owners of generation capacity "to reveal their plans for new construction and retirement of existing units any sooner than they have to" (Pansini 2004, 108).

The development of competitive markets in the United States during the 1990s prompted the Federal Energy Regulatory Commission to emphasize the importance of transmission planning on a regional basis. This emphasis led to the creation of regional transmission organizations, with planning (including expansion planning) as one of their eight minimum required functions.[19] However, traditional approaches to planning proved difficult to change because of the vested interests and resistance of the incumbent utilities and the state regulators who traditionally supervised the electric supply industry.

Inadequate planning is likely to be more costly for China because of the rapid growth of the power subsector. To prevent deterioration of the reliability of electricity supply during the transition period from command and control to more market-oriented management of the power industry, China urgently needs to review planning procedures at the generation, transmission, and distribution levels. Although it will take time to reform the project approval process, investment decisions for the five-year plans need to be based on a longer-term vision of the development of the sector. In particular, these plans should integrate structural changes of the load curves, demand-side alternatives to building new capacity, and technological innovation, in order to meet the demand at the least cost and in an environmentally acceptable manner. Reliability criteria should be based on the specific conditions of the different grids and on consumer willingness to pay the costs to meet them. This activity requires the involvement of and coordination among all parties concerned, including utilities at national, regional, and provincial levels. This coordination should include institutes and agencies involved in the development of engineering and reliability standards as well as the planning, environmental, and regulatory agencies.

Better maintenance procedures and increased reliance on preventive maintenance are essential to the dependability of the systems. The coordination of maintenance schedules and procedures also could contribute to increasing the reliability of supply. Furthermore, utilities should devise plans for rapid restoration of service in case of equipment failure. These plans, including emergency allocation schemes, need to be discussed with the authorities concerned to enable them to understand the system constraints and avoid undue interference in case of local interruptions of service. Assistance and mutual support among separate grids need to be discussed and agreed on to avoid domino effects and minimize supply interruptions in case of major blackouts. Most of these measures already exist; they need to be coordinated and codified to improve the reliability of China's power supply.

Improving the Security of Power Systems

As discussed earlier, the need to ensure the physical security of power installations has emerged since the September 11, 2001, attacks in the United States. "Electric service has become a necessity not only in the lives of individuals, but also in the operation of public services; e.g., water supply, sewage disposal, communications, transportation, health activities" (Pansini 2004, p. xiii). Many industrial countries have developed protection measures for the most sensitive installations, such as nuclear power plants. The United States and other countries are considering further security measures to reduce the vulnerability of power systems to attack. These measures include extending protection to installations that are critical to the integrity of the system, sectionalizing transmission systems into separate entities to avoid domino effects and extensive blackouts in case of failure or damage to parts of the system, and reducing the visibility of power installations.

It is clear that protection of the installations, changes in the design of power systems, and development of measures to minimize restoration time will increase the construction and operating costs of power systems. Experience in countries affected by political disturbances indicates that substantial increases in project costs have resulted with the transfer of security operation for installations (not specifically power or energy) to investors and project operators. "The usual economic considerations may have to be strained to new limits, and in some instances ignored" (Pansini 2004, x).

What is true for energy supply systems in general is also true for the electricity industry: security of supply can be improved by demand-side options that make the best use of available capacity, by geographic and fuel-type diversity of supply, and by improved system planning. These means to improve the security of power systems are simply provided for reference purposes. They could more properly be considered aspects of integrated resource planning, power system design, and electricity demand management.

As for oil products, emergency electricity allocation schemes should be considered—to determine priority allocations in times of scarce supply and to prevent collapse of the system and ensure that vital services have adequate electricity supply.

Choosing the Right Mix and Amount of Energy Supply Security Measures

Policy makers will have to decide, for each source of energy, what is likely to be the most effective mix of measures to enhance the security of supply. Some options to choose from have been provided in this chapter (see also box 5.6 on pp. 120–21). Each set of mixes will need to be reviewed from time to time as national and international circumstances change.

Policy makers will also have to decide what amount of security they should aim to buy on behalf of consumers. For example, the likelihood is that a useful degree of security against a short-term partial oil import disruption can be achieved by a relatively small investment in such measures as stockpiling and emergency allocation. To achieve a similar degree of security against the extreme and unlikely case of total, long-term disruption of oil imports would require huge investments. Beyond a point, security investments may yield diminishing returns and at the same time face declining acceptance in the society and a decreasing ability of the economy to pay for security measures.

It is not possible to arrive at a quantitative measurement of security of energy supply. Instead, against the background of assessments of the changing risks to energy security, tradeoffs will have to be made between security costs and benefits. The result of successful tradeoffs will be the establishment of a comfort zone within which the value of enhanced security approximates the cost of providing it and the relationship of these two variables is stable.[20]

Box 5.6

Strengthening China's Energy Security: China's Special Concern

Rising oil imports mean that China's oil supply will be out of China's political control. Policy measures are necessary to strengthen China's energy security.

Measures to Strengthen Long-Term Energy Security

China should foster national energy resources development with an emphasis on clean energy (natural gas and renewables) and on more reliance on market forces. These measures will help strengthen the country's energy security in the long term:

- Encourage development of national oil and gas resources through improving access to undeveloped resources, rationalizing fiscal terms for their development, and clarifying the role of national and international oil companies.
- Secure imports through diplomatic and trade relationships (with the Middle East for oil and with Russia for gas).

Measures to Strengthen Short-Term Oil Security

China also needs to strengthen its oil security in the short term. It can do so through these measures:

- Maintain oil stocks above minimum commercial requirements by developing a strategic petroleum reserve, mandating minimum commercial stocks, maintaining spare domestic oil production capacity (an assessment of domestic industry's capacity is needed), and providing military protection for oil traffic in the chokepoints of international tanker lanes.
- Build flexibility into the oil refining industry.
- Maintain or develop fuel-switching capabilities in large consuming industries.
- Set up emergency allocation schemes.
- Establish rationing systems for oil products.
- Increase international cooperation at global, regional, and bilateral levels.

At present, flaws in the power planning system are causing bottlenecks. Deregulation of the power industry will lead to California syndrome concerns.

Options for Better Security of Electric Power Supply

Several alternatives exist for better security of electricity:

- Improve planning and maintenance procedures.
- Integrate resource planning and implement demand-side management.
- Improve physical security of power systems by identifying and protecting critical infrastructure.

Overall Approach to Strengthening Security

An overall approach to strengthening security should begin with these steps:

- Assess the full range of security policy measures.
- Select the most cost-effective mix for the conditions of the time.
- Revisit the chosen mix of measures as national and international circumstances change.

Source: Authors.

Notes

1. Economic growth in the industrial countries (those of Western Europe, as well as Japan and the United States) was significantly impaired by the energy crises of 1973 to 1974 and 1979 to 1980. Countries were affected by energy shortages, supply uncertainties, inflation resulting from energy price spikes, and arguably by governments' mismanagement of their national energy economies.
2. One assessment (Taylor and Van Doren 2005) finds that the cost of the nearly 700 million barrels in the U.S. Strategic Petroleum Reserve over its 30-year life, allowing for inflation and a modest return on investment, has been US$41 billion to US$51 billion—between US$65 and US$80 on a per barrel basis. The oil currently in storage had a purchase cost much less than the present world price. Yet on many occasions, the oil stored has cost much more than the spot price for the corresponding grade. For similar views, see publications of the Cato Institute at http://www.cato.org/pub_cat_display.php?pub_cat=2.

3. There have been many public comments on the dispute. One of them is by the well-regarded Oxford Institute for Energy Studies (Stern 2006).
4. From 1985 onward, natural gas consumption in the United States recovered from the depressed levels of the era of excessive regulation, but it has been flat for several years.
5. The Bonn Conference on Renewable Energy gave rise to an international policy declaration on renewable energy, an international action program, and a set of policy recommendations addressed to governments. See http://www.renewables2004.de/en/2004/outcome.asp.
6. The announcement of the scale-up program can be found at http://web.worldbank.org/WBSITE/EXTERNAL/COUNTRIES/EASTASIAPACIFICEXT/EXTEAPREGTOPENVIRONMENT/0,,contentMDK:20545709~menuPK:502932~pagePK:34004173~piPK:34003707~theSitePK:502886,00.html. Details in Chinese are at http://www.worldbank.org.cn/Chinese/Projects/projects.htm.
7. The *Wall Street Journal* on March 7, 2006, reported remarks by Mr. Ma Kai of NDRC to the effect that, although the first of four strategic oil storage facilities had been completed, fill operations were going to be deferred to the end of 2006 in view of the current high price of international oil supplies (Dean 2006).
8. Global steel prices have, however, been declining since the summer of 2005.
9. The reserve is under the authority of the president of the United States. In September 2005, President Bush authorized the release of U.S. Strategic Petroleum Reserve crude oil to help overcome shortages caused by damage to production facilities during Hurricane Katrina. The volumes actually released appear to have been small—about 11 million barrels out of a total reserve of some 688 million.
10. It should be noted that the Suez Canal is another potential chokepoint. No significant flow of oil to China comes through the canal. However, the canal is of global significance and therefore of importance to China's security of oil supply, because an interruption of oil flow through Suez Canal would create a worldwide tanker shortage that would affect China.
11. A paper published by the South Asia Analysis Group (Kuppuswamy 2004) states "The [security of the] Straits of Malacca is an international issue, and safe passage has to be guaranteed under international

law. Hence it is not correct to say that only the littoral states have the right to deal with issues concerning the security of this waterway."
12. The IEA is the institutional vehicle for implementation of the 1974 Agreement on an International Energy Program. The Emergency Oil-Sharing System is embedded in the International Energy Program.
13. This quote is from the Preamble to the Agreement on an International Energy Program, dated November 18, 1974. The full text of the agreement can be found at http://www.iea.org/Textbase/about/IEP.PDF.
14. The basic IEA position on energy policy is contained in the document "Shared Goals," which was agreed to in June 1993 by the IEA ministers (IEA 1993). The document states, "Improved energy efficiency can promote both environmental protection and energy security in a cost-effective manner" and goes on to relate achievement of these objectives to "Continued research, development, and market deployment of new and improved energy technologies" that can "make a critical contribution to achieving the objectives outlined." Although the document is now 13 years old, it still forms the basis of the IEA's collective energy policies.
15. APEC's research arm, the Asia Pacific Energy Research Center, has provided background advice to APEC ministers in a report titled "Energy Security Initiative: Emergency Oil Stocks as an Option to Respond to Oil Supply Disruptions" (Asia Pacific Energy Research Center 2002). The report concluded that emergency oil stocks can contribute usefully to supply security and recommended that emergency oil stocks be separated from commercial ones, that the costs of emergency stocks be internalized in oil products prices (rather than subsidized), and that APEC members coordinate policy. APEC has not yet made any specific policy decisions on oil supply security measures. APEC's Energy Working Group is researching alternative ways to plan for regional emergencies or natural disasters such as the 2004 tsunami.
16. ASEAN member countries are Brunei Darussalam, Cambodia, Indonesia, the Lao People's Democratic Republic, Malaysia, Myanmar, the Philippines, Singapore, Thailand, and Vietnam. See http://www.aseansec.org.
17. See http://www.encharter.org.
18. In this section, a differentiation is made between the reliability of electricity supply and the security of electricity supply based on the

definition provided in Pansini (2004, xii): "[R]eliability encompasses contingencies resulting from flaws in design and human errors; security tends to negate damage and destruction of property and injury or death to humans deliberately caused by other humans."

19. The eight minimum functions of regional transmission organizations, as prescribed by the Federal Energy Regulatory Commission, are transmission tariff administration and design; congestion management; parallel path flow; ancillary services provider of last resort; administration of open access same-time information systems; market monitoring; planning, including of expansions; and interregional coordination.

20. The European Union provides an example of change over time in the security comfort zone for oil supply. For 30 years (1968–98), it was satisfied with a requirement that member countries hold stocks equivalent to 65 days worth of consumption of the major oil products. In 1998, this stockholding requirement was increased to 90 days. Currently, an increase to 120 days is being considered.

References

Asia Pacific Energy Research Center. 2002. "Energy Security Initiative: Emergency Oil Stocks as an Option to Respond to Oil Supply Disruptions." Asia Pacific Energy Research Center, Tokyo.

BP Global. 2005. *Statistical Review of World Energy 2005*. London: BP Global.

Dean, Jason. 2006. "China's Planners Slow Timetable for Oil Reserves." *Wall Street Journal*, March 7.

IEA (International Energy Agency). 1993. "Shared Goals." Adopted in a meeting of the IEA ministers, Paris, June 4. http://www.iea.org/Textbase/about/sharedgoals.htm.

Kuppuswamy, C. S. 2004. "Straits of Malacca: Security Implications." Paper 1033. South Asia Analysis Group, Noida, India. http://www.saag.org/papers11/paper1033.html.

Pansini, Anthony J. 2004. *Transmission Line Reliability and Security*. Lilburn, GA: Fairmont Press.

Schwartz, Peter. 1996. *The Art of the Long View*. New York: Currency Doubleday.

Shell Group. 2005. "Visions of the Future: Shell Launches New Global Scenarios Looking Forward to 2025." Press release, June 6. http://www.shell

.com/home/Framework?siteId=media-en&FC2=/media-en/html/iwgen/news_and_library/press_releases/2005/zzz_lhn.html&FC3=/media-en/html/iwgen/news_and_library/press_releases/2005/global_scenarios_launch_06062005.html.

Stern, Jonathan. 2006. "The Russian Ukrainian Gas Crisis of January 2006." *Oil, Gas, and Energy Law Intelligence* (January): 1–17.

Taylor, Jerry, and Peter Van Doren. 2005. "The Case against the Strategic Petroleum Reserve." Policy Analysis 555, Cato Institute, Washington, DC.

World Bank. 2004. *Toward a Sustainable Coal Sector in China*. Washington, DC: World Bank.

World Bank and IESM (Institute for Economic System and Management). 2000. "Annex 2: A Comparative Analysis of the Upstream Fiscal Terms for China." In *Modernizing China's Oil and Gas Sector: Structure, Reform, and Regulation*. http://www.worldbank.org.cn/English/content/oil-gas.pdf.

CHAPTER 6

Getting the Fundamentals Right

Key Messages

Appropriate market-based reforms, particularly sound price-setting methodologies and tools, are fundamental to an efficient, sustainable, and secure energy sector. China has made substantial progress in gradually freeing up energy prices, relaxing controls, and even, in some instances, allowing market forces to determine them. However, further reforms are needed to foster competition in energy commodity markets, to better determine and regulate prices in the monopolistic segments of the electricity and gas supply chains, to progressively reflect the environmental effects of energy use in costs and prices, and to use fuel taxation as an instrument of energy policy.

With respect to energy commodities—natural gas, crude oil, refined petroleum products, coal, and electricity—prices should, in principle, be set in a competitive market through the interplay of buyers and sellers. Competition can be fostered by freeing energy commodities from remaining price controls and opening the country's markets to more domestic and foreign energy producers, sellers, and buyers. Substantial progress has been made, but further action is needed. Coal prices are the closest to reflecting competitive market conditions, as a result of the recent complete liberalization of prices and contractual negotiations. Price controls on crude oil and oil products can lead to suboptimal use of resources, financial damage to the refining industry, and economic losses from smuggling. Controls should therefore be gradually lifted to allow domestic prices to fully reflect both international market prices and supply and demand conditions in China. Control of maximum final consumer prices should be retained until workable competition is achieved. Then, as conditions for market entry and exit from the different segments of the oil products market improve, regulation can give way to monitoring.

When a competitive wholesale market in natural gas is developed, prices of the natural gas commodity will be determined in contract negotiations between suppliers and buyers, including distribution companies, supplemented by a competitive spot market. The gas commodity price will be substantially unregulated and unbundled from transportation and distribution tariffs. The natural gas subsector reform agenda, including issues related to market structure, pricing, and a transition schedule, should be clarified in a policy statement and promptly implemented.

Pricing policies for natural monopolies on energy transmission and distribution services in China do not currently provide adequate revenue relative to financial needs, especially for gas and power, thus impeding consumers' access to suppliers and slowing reform. The prices and services offered by gas and oil pipelines and by power supply networks should be regulated to shield consumers from monopoly abuses. The State Electricity Regulatory Commission (SERC) should be comprehensively empowered and given the proper resources to carry out its mandate. A modern downstream gas regulator should be established, either separately or as an autonomous department of SERC.

As for adverse environmental effects of energy, mechanisms must be put in place to ensure that the effects of production, transformation, and use of energy are reflected in producers' costs and consumers' prices. Essentially, these mechanisms fall into one of two categories: regulatory mechanisms (in which the state dictates standards to be met) and market-based mechanisms (which involve the use of taxation or other mechanisms to price environmental costs into producers' and consumers' decision making). The choice of instruments will depend on many factors—political, economic, and administrative—and will require extensive analysis and consultation with all relevant players. Whatever the nature of the measures adopted, there will be an impact, direct or indirect, on the price of energy commodities.

Fuel taxation in China is very low both in absolute terms and relative to other countries. The government should consider levying higher taxes on fuel use to promote energy conservation, raise revenue for general government purposes, and promote the development and use of environmentally friendly technologies.

Market mechanisms for energy commodity pricing, modern regulation of natural monopolies, pricing-in of environmental effects, and modern fuel taxation can together contribute importantly to China's energy sustainability.

Furthering the Reform Agenda and Developing a Sound Pricing Framework

China is confronting the imperative of attracting and maintaining a huge level of investment in the energy sector to progress toward its growth objectives in an environmentally sustainable manner. Attracting investment requires incentives for efficiency in production, system operation, and consumption if the greatest benefits to consumers and to the economy are to be realized. However, efforts initiated by the government to unbundle the oil, gas, and electricity supply chains and to introduce competition raised some concerns about

- The dissipation of efficiencies that may presently exist as a result of vertical integration
- The security of supply and the integrity of the power networks as a result of the unbundling of the power industry
- The impacts on consumers—particularly those on low or fixed incomes—of possible increases in energy prices that may be required if the energy industry is to be properly financed to support its needs for expansion and if environmental security is to be achieved.

These challenges and concerns, however, are not unique to China. Over the past 30 years (and in some cases longer), policy makers in many other countries have successfully wrestled with these challenges and concerns. After painful trial and error, competitive pricing with targeted and performance-based subsidies became the central feature of energy policy in almost all industrial and in some advanced developing countries.

Economic principles and international practice show that when prices are formed by competitive forces operating in a free market, they tend to ensure efficient supply and limit consumers' demands to efficient levels. Efficient levels of demand are essential to sustainable development. Competitive prices help achieve economically efficient use of resources by ensuring that the costs of meeting consumers' needs are sufficient to pay for the private resources used in meeting those needs.[1] In a competitive economic environment, market prices continuously signal to producers the preferences and willingness to pay of energy consumers, and they signal to consumers the prices that producers require to put a particular volume of the commodity on the market.

Competitively determined prices can be relied on to continuously balance supply and demand without (or with very limited and targeted) government interventions. When prices are set by competition, markets will always clear, reducing notably the risks of shortages and surpluses as well as their negative economic consequences. Competition and market prices have already brought about dramatic beneficial changes in many other sectors of China's economy. However, competitive forces by themselves are unlikely to be sufficient to ensure environmental sustainability. Appropriate taxation or other policies, including those discussed in chapter 4, are needed to achieve this objective. Market mechanisms, augmented by appropriate policies to deal with external costs (such as the costs of environmental degradation) and with social costs (such as the impact of external costs on the poor), ensure efficient use of capital, labor, and natural resources such as oil and gas deposits; improve the performance of the energy sector; and help the overall economy to achieve the goal of becoming a resource-saving society.

After about two decades of remarkable progress, the pace of reform of China's energy industry has slowed during the 10th Five-Year Plan (2001–05), and the future direction seems uncertain. The consensus that allowed the early changes (especially at the enterprise level) to be implemented rapidly and successfully is weakening. Divergent views have emerged among government agencies, and narrow corporate interests seem to have taken precedence over achievement of the government's reform objectives. Moreover, government agencies have realized that further market reforms would fundamentally change or reduce their roles in the sector. The resulting conflicts of interest seem to have slowed (if not stalled) reform of the energy (particularly power and gas) sector during the 10th Five-Year Plan. Streamlining and deepening the reform process are essential to successfully addressing the challenges facing the energy sector.

Despite the progress achieved during the past 20 years or so, China's energy sector now faces both structural and operational problems, which, if not resolved, will severely constrain its development and its ability to meet the growing energy needs of the economy. Reviving the reform process, emphasizing further reliance on market mechanisms, and encouraging competition are vital to improving the efficiency of the sector and securing future energy supply.

This chapter outlines specific priorities for the oil, gas, coal, and electricity subsectors. The most urgent general priorities for the energy sector are to speed up the restructuring of the sector, to develop sound pricing policies to enhance competition, to empower large energy consumers who have so far had little say in the reform process, and to develop policies designed to promote the environmental sustainability of energy production and use. Delaying these actions will only strengthen the vested corporate and local interests that helped stall the reform process. Piloting and gradual approaches are not needed to empower consumers. The experience of the past two decades indicates that Chinese companies have a great capacity to adapt to the market and that most of them would welcome the opportunity to negotiate longer-term contracts with producers and suppliers.

Energy Pricing in Competitive Markets

Appropriate price-setting mechanisms are fundamental to achieving an efficient, sustainable, and secure energy sector. China's energy sector went through dramatic structural changes during the past two decades: separation of government and enterprise functions, separation of state-owned enterprises' core businesses from other businesses and social functions, transfer of these core businesses to newly incorporated companies, partial stock market listing of several large energy entities, full liberalization of coal prices, and increase of most energy prices to average costs of supply—especially in the power and oil products subsectors. Until the recent surge in world oil prices, oil product prices were linked, with a lag, to Singapore market prices. Energy pricing policies have, across the spectrum, been moving in the direction of market opening and greater reliance on competitive pricing. However, the changes have been piecemeal and comparatively slow. The energy sector has now reached a point where further progress toward achieving the government's development and reform objectives requires fundamental and comprehensive policy decisions on energy pricing, including whether and how prices are to reflect the existence of external costs and the appropriate role of energy commodity taxation. This task is not an easy one, but efficient energy pricing is one essential, foundational element of a successful, sustainable modern energy policy.

In the absence of externalities, efficient pricing exists when the incremental cost of producing, transporting, and distributing a commodity is equal to the price at which it is sold—its *market price*. The price of the commodity is efficient at this level because the value of the commodity to the consumer is equal to or greater than the cost of producing it (where the cost of production includes a reasonable rate of return to the producer). This result will occur without government intervention when workable competition exists—that is, when the number of sellers and buyers (including international companies trading with enterprises in China) is reasonably large, so that no single buyer or seller can, by itself, influence the price of the commodity; when information about prices and quantities is readily available; and when the costs of conducting transactions are reasonably low.

With respect to energy commodities—natural gas, crude oil, refined petroleum products, coal, and electricity—the government has issued several regulations designed to promote competition and open China's energy markets to more sellers and buyers, including more domestic and foreign producers. Current energy pricing is, however, still not fully conducive to optimum resource allocation. Take three examples:

- *Natural gas.* Natural gas prices to some consumers, particularly fertilizer manufacturers, are at present lower than required to cover all the costs incurred in the production, transportation, and distribution of this fuel. As a consequence, natural gas consumers are not confronted with the true costs of their decisions to use this resource. This fact, in turn, implies that investors somewhere in the chain of production, transmission, and distribution are receiving less than their costs of production. There are two salient consequences of this situation: first, less gas will be supplied into the market, and second, some of the gas that is supplied will be used in ways that are economically wasteful. Incorrect signals are transmitted to both producers and consumers. If the

government has policy goals for specific industries (in this case, supporting domestic fertilizer production), it is undoubtedly more efficient to achieve those goals through direct subsidies to the sectors in question than by distorting energy prices across the board.
- *Coal.* There was a failure in recent years to pass electricity price increases stemming from higher coal prices through to final consumers on a timely basis, especially during periods of power shortage. As a result, consumers were not encouraged to reduce nonessential consumption. This factor, in turn, diverted electricity consumption away from the sectors of the economy that valued it most. Again, the result was that consumers received incorrect signals.
- *Electricity.* The current generation contracts based on an energy (kilowatt-hour) price and a minimum number of operation hours distort the dispatch of generating units and significantly reduce the efficiency of the power subsector. As a result, incorrect signals are sent to producers.

The inevitable, growing integration of the oil and gas subsectors with regional and global markets will increase the distortions created by government-controlled commodity prices, because import prices are negotiated on the basis of international and regional market conditions. If domestic prices are not aligned with going market prices, subsidies will be required (when domestic prices are lower than market prices) or surpluses will be generated (when domestic prices are higher than market prices) that could put heavy burdens on government budgets, in the first case, or on consumers, in the second.[2]

Setting Sound Regulated Tariffs

With respect to the natural monopoly components of the power industry, a comprehensive policy needs to be developed and issued as soon as possible to speed up the design of regulated tariffs that are conducive to efficiency.[3] Market pricing generally cannot work for the network services (pipelines and transmission and distribution systems) that deliver energy commodities, because they are natural monopolies.[4] Regulation of tariffs and conditions for access to network services is necessary to substitute for competition and ensure the proper functioning of commodity markets. Regulated tariffs should be fair to suppliers, allowing them to recover their costs while shielding consumers from monopolistic abuses. At the same time, incentives for continued efficiency in investment and operations need to be provided.

Pricing policies for energy transmission and distribution services that are natural monopolies in China do not currently provide adequate revenue relative to financial needs, especially for gas and power. This problem is impeding consumers' access to gas suppliers and electricity generators, slowing the reform processes pursued by the government, and eroding the capacity of the newly unbundled power grid and gas pipeline companies to provide transmission services.[5] It is critical, therefore, that modern regulatory institutions be established and that sound principles and practices of regulation be implemented.

The cardinal principle of price regulation of natural monopolies is to base regulated prices on efficient costs. If service providers were reimbursed for their actual costs, they would have no incentive to improve efficiency. This is the basic conundrum of regulation: how to base prices on cost while retaining incentives for efficiency in operations and investments.

Several studies carried out in China showed that the current gas and power tariff systems do not meet these principles (World Bank 1997, 2001). Tariffs are rudimentary and are not set up to cover costs systematically (although the average price level is probably not far from average costs). Developing sound tariffs for power and gas transmission and distribution is essential to furthering competition in the commodity markets.

The principles and objectives of regulated tariff design for power and gas transmission and distribution should focus on achieving economic efficiency without unjustified discrimination in the provision of services, while ensuring that revenues generated are sufficient to recover costs that are prudently incurred. At the same time, tariffs need to be easily understandable, transparent, predictable, and adaptable. Methods used in different countries have evolved over time, as described in appendix I. There is a consensus on the principles of tariff design, although differences exist in the approaches used to attain them. The three steps to the design and setting of a tariff are as follows:

1. Determine the total revenue requirement in the base year—that is, the amount of money required to cover the costs of operating the facilities in question (transmission pipeline or power line, distribution system for gas or electricity), including depreciation, taxes, and a fair return on investment.
2. Develop a schedule of rates (prices) for the services to be provided to users of the facilities at levels that will generate the revenue requirement under certain assumptions about the use that will be made of those services.
3. Provide a method of adjusting rates from year to year.

For each of these steps, certain rules apply that are also the subject of consensus. These principles and the rules governing their application are discussed in more detail in appendix I.

The following principles should apply when setting regulated tariffs:

- Tariffs to final consumers should cover all the costs of efficient provision of regulated services, meeting predefined targets for quality of supply and service.
- Tariffs should provide a return on—and recover depreciation of—all the investments that the regulated company needs to incur for the efficient provision of its services.
- Investments should be amortized over their full useful lives. (Very short amortization periods are a major distorting factor in China; in the regulated industries they tend to return the capital investment too quickly, resulting in excessively high tariffs in the early years and excessively low tariffs when the investment has been recovered.)
- Tariffs should be as close to marginal costs as possible.
- Tariffs should offer incentives for operational efficiency.
- Efficiency must be understood as the lowest tariffs consistent with certain obligations, such as service quality and network expansion.
- The method of regulation should not have the perverse effect of discouraging energy-efficiency programs, which occurs when regulation denies the service provider all or part of the revenues associated with a given energy-use saving.
- If subsidies are deemed to be desirable, for example, to enable the purchase of electricity or gas to meet basic needs (as for low-income consumers) or to protect particular consumer groups (such as small-scale farmers), they should be designed to minimize the distortion in price signals and fully funded by government to prevent adverse impacts on the service providers.

Regulated companies must adopt a proper system of accounts and accounting procedures that provide detailed and accurate financial, cost, and consumption data. In addition, the regulated entities must develop proper techniques to measure throughput losses; supply interruptions; voltage (or pressure, for gas) and frequency variations; and other parameters of supply. These data lie at the heart of the SERC's ability to regulate the tariffs and service offerings of all companies effectively, taking into account factors that encourage economic use of resources, good performance, and optimum investments. As part of the full tariff filing, companies should provide the following data and reports:

- Proposed aggregate revenue requirement, with detailed support for each element[6]
- Expected revenue at the prevailing tariffs
- Proposed tariffs, terms, and conditions
- Expected revenue from the proposed tariffs
- Financial accounting and any other information as required by the regulator to properly situate the applied-for revenue requirement and tariffs in the context of the company's overall operations and financial situation.

Taxing Energy Commodities

The level of taxation of energy commodities in China is very low both in absolute terms and relative to that of other countries. Taxes may be imposed for a variety of reasons: to raise revenue for the state, to pay for infrastructure such as roads, or to encourage energy conservation. As discussed below, taxes are also increasingly viewed as a means of incorporating in energy prices the environmental costs associated with the production and use of energy.

Comparable international data are often hard to find, and comparisons are difficult to make. However, one recent comprehensive study (Metschies 2005) of fuel prices and taxation proposed two indicators of the extent of taxation of oil products for a large number of countries:

- The fuel tax contribution to total state revenues
- The retail prices of oil products in various countries compared with what the price would be if taxation were very low.

Table 6.1 shows the contribution of fuel taxes to total government revenues for a number of countries, including China.

Table 6.1 is helpful but necessarily gives only an imperfect indication of the fiscal balance for fuel taxation because it reflects individual countries' choices about the total tax burden. Those choices derive from different decisions about the appropriate size of the public sector of the economy. The ratio also reflects differences in the mix of commodity and income taxation across countries. For example, although the contribution of fuel taxation to total state revenues is the same in France and the United States, this does not mean that the levels of fuel taxation are roughly comparable. Clearly, as is evident in table 6.1, the level of fuel taxation is much higher in France, but so is the overall taxation level.

Table 6.1 Fuel Taxation in Selected Countries

Fuel tax contribution to total government revenue		Retail price of superoctane gasoline		
Country	Percentage[a]	Category	Country	Price[b] (U.S. cents/liter)
Venezuela, R. B. de	−17	Fuel subsidization	Venezuela, R. B. de	4
Algeria	−5		Algeria	32
China	0		China	48
Brazil	5	Moderate fuel tax	United States	54
United States	12		Brazil	84
France	12		India	87
India	15	Very high fuel tax	Japan	126
Japan	17		Korea, Rep. of	135
Korea, Rep. of	33		France	142

Source: Authors.

a. Negative numbers indicate that the use of oil products is being subsidized in the country.

b. Prices are for November 2004.

Table 6.1 focuses on the price of one oil product—superoctane (premium) gasoline—expressed in U.S. cents per liter, across a number of countries. Although the table does not show the tax level directly, country data can be compared with a benchmark designed to measure what the price would be in the absence of taxes. The benchmark chosen was the retail price of gasoline in the United States, where taxes are low. Thus, those countries with gasoline prices below those of the United States can be characterized as subsidizing fuel, and those with higher prices as having fuel taxation on a net basis.

Given the data problems, these numbers can at best give only a rough indication of the level of taxation across countries. It is evident, however, that fuel taxation in China is very low compared with other industrial countries (especially Japan and the Republic of Korea) that are large net oil importers; indeed, it appears to be close to zero on a net basis. There is clearly room to levy higher taxes on fuel use to promote energy conservation, to raise revenue for general government purposes, or to subsidize the development and use of environmentally friendly technologies. Some of these objectives are contradictory; for example, the greater the conservation effect, the smaller would be the amount of revenue raised. In the near term, the revenue effect can be expected to dominate, because the scope for fuel savings will be limited by the technology embodied in the existing stock of vehicles. Over time, however, higher energy prices will induce companies and individuals to purchase more fuel-efficient units. Thus, the conservation effect will increase, and the revenue effect decline. In China, where the current contribution to the government budget is very low and conservation a very high priority, tax increases could meet both objectives.

Measures to Mitigate the Environmental Effects of Energy Use

Many industrial countries have begun to adopt measures—including the taxation of energy commodities—to reduce the environmental damage associated with energy use. They are doing so even though their environmental pollution problems are, in many cases, much less acute than China's. The challenge is a difficult one but must be faced and progressively addressed to meet the goal of sustainability for the energy sector. The use of appropriate measures, including taxation, to align energy prices more closely with the total costs, commercial and social, of its production and use can be important complements to the policies described in chapter 4 to put the sector on a sustainable growth path.

As noted, market pricing is efficient as long as markets are reasonably competitive. However, efficiency also requires two other conditions. First, all the costs associated with production should be borne by producers. (If costs are not borne by producers, they will not, of course, be reflected in the market price.) Second, all benefits to consumers resulting from their use of the commodity should be reflected in their willingness to pay for it. Such costs and benefits are referred to as *externalities* because they are not reflected in the market prices of the associated commodities.

The concept of externality is most often used in the context of negative environmental externalities, such as air pollution, which damages human health, farm crops, natural vegetation, and even building structures and from which the polluter may not suffer direct or indirect damages. In principle, externalities may also be positive, as for example, in the case of a bee farmer whose bees help pollinate the fruit trees of a nearby orchard. Essential to the definition given in the previous paragraph are both the lack of participation in the decision concerning the economic activity by one or more of the parties affected and the absence of full compensation of the costs or benefits accruing to the receiving party. It should be noted that under this definition, environmental pollution might conceivably not be an externality if those who suffer from the negative impacts of that pollution are fully compensated (Virdis 2001).

Environmental security is a critical component of energy sustainability in China. Thus, an essential element of China's energy policy must be the development of mechanisms to mitigate environmental damage. There are many possible ways of devising such mechanisms, but most fall into two categories, which can be characterized as *regulatory* and *market based*:

- *Regulatory instruments* are those in which the state dictates standards to be met in the production or use of commodities—for example, standards for allowable sulfur dioxide emissions from power stations—and those that are reflected in producers' and consumers' costs—for example, in terms of the capital and operating costs of flue gas desulfurization. When regulation is used, producers or users affected by the standard face the choice only of accepting the costs of conforming to the standard or of not producing or using the energy in question. The impact on polluting emissions is predetermined.
- *Market-based instruments* involve the incorporation in prices of externalities, such as the adverse effects of sulfur dioxide emissions, through taxation or through mechanisms such as tradable permits, so that decisions about energy production and use—and their environmental consequences—are made by producers and consumers on the basis of price signals. When taxation is used, the market's response to the resulting higher prices will determine the influence on pollution through the decisions of buyers and sellers. The state may then adjust the level of taxation to achieve its pollution reduction goals. In the case of tradable permits, the state predetermines the allowable total amount of emissions but effectively allows that level to be achieved at least cost through a market in emissions permits. In general, the use of market mechanisms leaves it to market participants to determine least costly ways of achieving environmental goals.

The choice among different policy instruments is not always easy. As one analyst has put it,

> There exists a vast economic literature on the relative efficiency of various policy approaches to address externality problems, and on the merits of price and market instruments . . . versus [regulatory] instruments (prohibitions, standards, directives, etc.). Following this debate, governments in the last decade have changed their environmental policy portfolio towards more market-oriented and decentralized approaches. The debate, however, is not over, as in many

situations insufficient or asymmetric information, high transaction or policy implementation costs, or the social hazards of some activities leave little practical room for policy choice. Although inefficient in many situations and difficult to enforce, mandatory standards are sometimes the only viable solution. Furthermore, policy making often has to serve and reconcile multiple objectives, not only health and environmental protection. (Virdis 2001, 208)

Table 6.2 illustrates the nature of market-based instruments and regulatory measures and classifies them as to whether they act directly or indirectly on the environmental problem.

Each type of instrument and incentive has its advantages and disadvantages. A substantial virtue of market-based incentives is that, in principle, they "provide increased flexibility for the targeted community, enabling it to make compliance decisions that minimize compliance costs and thus maximize profits. According to environmental economic theory, the ability of [market-based incentives] to enable profit-maximizing behavior gives them a major advantage over traditional command-and-control [regulatory] approaches. [Market-based incentives] are also more attractive because, in theory, they reduce government implementation costs, raise government revenues, and reduce budgetary outlays, thus reducing the costs (both to government and industry) of meeting societal objectives" (National Roundtable on the Environment and the Economy 2005, 10).

The choice of instruments will depend on many factors—economic, political, and administrative—and will require extensive analysis and consultation with relevant interest groups. For example, the government may not have sufficient information to determine appropriate levels of taxation.[7] Further, the nature of some markets, such as that for rental housing, may be such that price signals either are not passed through to energy users (if rents are controlled) or are not acted on. They might

Table 6.2 A Classification of Market-Based and Regulatory Instruments

	Direct instruments	Indirect instruments
Market-based	Emissions fees Marketable permits	Environmental taxes (on motor vehicles, fuels, and polluting substances such as sulfur dioxide)
Regulatory	Emissions standards Energy-efficiency standards (buildings and vehicles)	Technology standards

Source: Adapted from Blackman and Harrington 1999.

not be acted on if tenants are unlikely to stay in a unit for a long enough period to recoup the benefits of investing in energy-saving technology. In such cases, regulatory measures may be appropriate. Whatever measures are adopted, there will be an impact, direct or indirect, on the price of energy commodities. If China is to have a sustainable, environmentally secure energy future, such measures need to be fully assessed, evaluated, and adopted as soon as possible.

Energy Commodity Price Policy: Status and Necessary Changes

Coal prices are the closest to reflecting competitive market conditions. The government began lifting price controls in 1993 by allowing producers to negotiate sale prices with users—except for the state power companies, which continued to benefit from controlled, relatively low, prices. Coal prices were liberalized in 2006 after being capped in 2005 with an 8 percent increase with only partial passthrough of that increase to electricity final users. This liberalization of prices and the contract negotiations between coal suppliers and power generators is bringing the coal industry closer to the ideal of a competitive commodity market.

The liberalization will improve the competitive environment in the subsector and encourage sellers and buyers to rely on long-term contracts, reducing price volatility and contributing to market stability. It is also critical that measures be developed and implemented to mitigate the environmental consequences of coal use. Removal of the price cap is consistent with the goal of environmental sustainability, but supplementary measures—either market based or regulatory—are likely to be necessary.

Price controls on crude oil and oil products should be gradually lifted to allow domestic prices to reflect both international market prices and supply and demand conditions in China. In 1993, China emerged as a net oil importer, and it has since become one of the major actors in regional and international oil markets. Furthermore, with the agreed progressive opening of downstream oil product markets to foreign operators from 1996 onward and with China's growing imports, the country's oil market will increasingly be influenced by and will influence regional and international markets. China is at a crossroads: it could either repeat the errors committed by so many countries in the past or learn from them and devise policies to facilitate its integration into regional and global markets, to reinforce its position as a major player in these markets, and to invest more effort in establishing innovative modes of cooperation with other major players to ensure a secure and long-term oil supply in the most economically efficient way (see appendix J, which presents lessons from international experience of unsound energy pricing).

The experience of most industrial countries and of developing countries comparable to China shows that the only way to achieve comprehensive market-clearing behavior (or self-balancing markets) for crude oil and oil products, where imports are a significant proportion of supply, is to allow and enable domestic markets for crude oil and oil products to fully reflect international trends. Other approaches would involve cumbersome state regulation of commerce and prices, with ensuing counterproductive effects on the reliability and quality of supply and the overall efficiency of the economy.

Crude oil and oil products markets are not characterized by natural monopoly (except for some pipelines). However, the existence of quasi-monopolistic or strong oligopolistic conditions in China's regional markets for oil products still requires control of maximum final consumer prices until workable competition is achieved. The policy to control these prices by linking them to the Singapore crude oil market price with 5 percent leeway for negotiations between oil companies and refineries and consumers brought several benefits. However, it has induced speculative behavior in recent years. It is important now to further open the subsector to new product and service providers and to promote competition among the three major national companies and new entrants. This development would lead to achieving a critical mass of competing suppliers, meeting the country's commitments under the World Trade Organization agreement, and developing a competitive market for oil products.

Consideration should also be given to increasing taxes on refined petroleum products. Such taxes would encourage conservation, thereby contributing to energy security and sustainability. Higher taxes and the associated prices have the incidental effect of reducing the shock of surges in world oil prices on consumers: the greater the tax on the end-use product, the smaller will be the percentage change in end-use prices associated with a given change in crude oil prices.

Development of a competitive market pricing system for natural gas requires further reform of the subsector. There are several indications that Chinese authorities are targeting the development of a competitive wholesale gas market.[8] The wholesale market model, already functioning in North America and under development in Europe, focuses on achieving workable competition in the bulk supply of gas to large industrial consumers, who are declared eligible to make their own gas purchase arrangements, and to the urban gas distribution companies (UGDCs).

It does not necessarily extend competitive choices to small consumers at the retail level—at least, initially. Small consumers would, of course, benefit from efficient gas purchasing by their UGDC suppliers.

When a competitive wholesale gas market is developed, prices of the commodity (natural gas) will be determined by negotiated contracts and by a competitive spot market for eligible industrial and commercial consumers and distribution companies. The gas commodity price would be substantially unregulated and unbundled from transportation and distribution tariffs. UGDCs would be allowed to pass commodity prices through to final smaller consumers, bundled with transmission and distribution tariffs. Special measures could be adopted to address the needs of low-income customers. More complex, competitively oriented changes in network pricing, such as tradable gas transmission rights, could be introduced at a later stage.

This transition to wholesale competition in the gas market has been researched in detail in studies carried out jointly by the World Bank and Chinese government agencies (World Bank and IESM 2002a, 2002b; World Bank and PetroChina 2003). The most urgent policy measure is to formalize the gas industry reform agenda and schedule in a statement similar to State Council Document No. 5 for the power subsector. The statement would clarify the envisioned market structure and establish a transition schedule along the lines set out in appendix K, which deals more generally with issues of gas price formation and gas sector reform.

The document should provide large consumers and provincial and municipal UGDCs with direct access to gas producers and traders (a prerequisite for efficient competition in the subsector). It should create an agency to provide modern regulation for the downstream gas industry and to monitor the development of the competitive market.

Access of large consumers to generators and development of sound regulated prices for transmission and distribution are essential to furthering the reform of the power subsector. In the power industry, Document No. 5 was a major step forward. It clearly stated that the objectives of the reform in China are to continue breaking up the monopolistic structure of the industry and gradually to expand competition to improve its efficiency and ultimately provide customers with the best service at the lowest possible cost. The plan details these objectives in eight steps:

1. Break up monopolies.
2. Introduce competition.
3. Increase efficiency.

4. Perfect pricing mechanisms.
5. Optimize resource allocation.
6. Promote industry development.
7. Promote the formation of a national grid.
8. Establish a competitive electricity market.

Appendix L outlines the main elements of the pricing system needed to support adequate implementation of the policy reform outlined in Document No. 5.

Document No. 5 clearly envisions that large customers will purchase directly from generators and that they will pay for the transmission and distribution services they receive. However, large customers' access to generators and effective competition is hampered by the slow and piecemeal change in pricing policy. The pricing policy outlined in Document No. 5 also implies that wholesale prices will fluctuate and should be passed on to end users. Consumer prices need, therefore, to fluctuate also. As for natural gas, special measures can be implemented to assist low-income customers.

Introducing effective competition, the primary objective of Document No. 5, requires

- Giving large consumers (eligible consumers) and distribution companies access to purchase freely from generators. As in the gas subsector, having a large number of generators and customers is a prerequisite to achieving workable competition.
- Establishing tariffs to be paid for the use of transmission networks by eligible customers and by distribution companies.
- Developing pool or spot-pricing mechanisms to set market prices in real time. These spot prices will be used to clear the imbalances that naturally occur between contracted amounts and the amounts actually generated for the contract or taken out of the grid by the consumer. The spot prices need to be integrated with the mechanism for giving competing generators and customers access to the networks. Competitive power markets have been successfully piloted at the provincial and regional levels in China.
- Establishing a method for pricing contracts between distribution companies and generators (perhaps through auctions, cost of service, or bilateral contracting).
- Establishing a method for retail tariffs that reflects the fluctuating prices of the commodity and includes transmission and distribution tariffs.

Finally, a sustainable policy requires that the environmental consequences of generation be acceptable. Various policy measures are available to achieve this goal, including the adjustment of prices to incorporate environmental costs. Wide consultation and discussion among all concerned Chinese agencies will be necessary to assess and determine the most appropriate measures.

Notes

1. Public resources may be used, as in environmental degradation, the costs of which are not borne by producers or consumers, absent appropriate public policies.
2. The *Wall Street Journal* of April 26, 2006, reported "Sinopec's net profit in 2005 rose 14 percent, to Y 40.92 billion—in part thanks to a Y 10 billion government cash injection to offset its refining losses [caused by government-controlled fuel prices that are below international levels]. By comparison, profit rose 28 percent at PetroChina and 57 percent at CNOOC" (Oster 2006).
3. World Bank work in this area includes a consulting report examining opportunities to improve the energy efficiency of China's power sector on the supply side and on the demand side through improvements in pricing and pricing methods. China's ongoing power crisis means that these are especially important and timely issues.
4. The network services are natural monopolies because large economies of scale and scope exist, and the unit costs of providing their services decline over a wide range of capacity. Thus, the minimum cost per unit of transportation exists only when such systems are very large. In summary, the goal of regulation is to ensure that the advantages of natural monopolies (tariffs reflecting their low costs) are realized and the disadvantages associated with their potential market power (discrimination, excess profits) are avoided.
5. *Unbundling*, in this context, means the separation of the pricing of markets for energy commodities (for example, electricity and natural gas) from the pricing and provision of delivery services for those commodities (for example, electricity transmission and gas distribution).
6. The elements of the revenue requirement are the operating and maintenance expenses of providing services, a fair return on investment in assets made by the company to provide those services (the *rate base*), depreciation of that investment, and taxes and other mandated costs.

7. The appropriate level of taxation will depend on the extent to which the state wishes to reduce emissions and on the way in which buyers and sellers of the energy commodity in question respond to the resulting higher price. That is, in technical terms it will depend on the price elasticities of supply and demand. Such information is not always easily obtainable; it requires detailed analysis of market behavior.

8. *China Daily* reported on December 27, 2005, that "The government has decided to phase out its current practice of pricing natural gas, with an aim to form a market-oriented price mechanism in the sector.... Starting next year, natural gas prices will be modified once every year, a spokesman said.... 'In the long run, natural gas prices should... be decided by the market, not the government,' said the spokesman. But he added that the government should introduce the reform gradually because state-owned PetroChina and Sinopec still dominate gas exploration, gas transportation, and sales on the Chinese mainland" (*China Daily* 2005).

References

Blackman, Allen, and Winston Harrington. 1999. "The Use of Economic Incentives in Developing Countries: Lessons from International Experience with Industrial Air Pollution." Discussion Paper 99-39, Resources for the Future, Washington, DC.

China Daily. 2005. "Natural Gas Prices to Be Market Driven." December 27. http://english.peopledaily.com.cn.

Metschies, Gerhard. 2005. *International Fuel Prices 2005*. 4th ed. Eschborn, Germany: Deutsche Gesellschaft für Technische Zusammenarbeit.

National Round Table on the Environment and the Economy. 2005. *Economic Instruments for Long-Term Reductions in Energy-Based Carbon Emissions*. Ottawa: National Round Table on the Environment and the Economy.

Oster, Shai. 2006. "Sinopec Is Pinched by Price Caps at Home as Global Oil Costs Rise." *Wall Street Journal*, April 26.

Virdis, Maria Rosa. 2001. "Energy Policy and Externalities: The Life-Cycle Analysis Approach." In *Externalities and Energy Policy: The Life-Cycle Analysis Approach*, 195–234. Paris: Nuclear Energy Agency and Organisation for Economic Co-operation and Development.

World Bank. 1997. "China: Power Sector Regulation in a Socialist Market Economy." Discussion Paper 361, World Bank, Washington, DC.

———. 2001. "Fostering Competition in China's Power Markets." Discussion Paper 416, World Bank, Washington, DC.

World Bank and IESM (Institute of Economic System and Management). 2002a. *China: Economic Regulation of Long-Distance Gas Transmission and Urban Gas Distribution.* Washington, DC: World Bank.

———. 2002b. *China: Regulatory Framework for Long-Distance Gas Transmission and Urban Gas Distribution.* Washington, DC: World Bank.

World Bank and PetroChina. 2003. *Gas Price Formation in China: Transmission Tariff Design.* Washington, DC: World Bank.

CHAPTER 7

Shaping the Future toward Sustainability

Key Messages

China's present energy trends are not sustainable: the road to sustainability requires a new legal framework and policies and programs to create and support a resource-conscious society. The government has recognized the need for change and has outlined four areas of policy emphasis to address it: focusing on efficiency, relying on competition, enforcing laws, and making energy a national priority by creating a national Energy Commission.

The energy law now in preparation provides the opportunity to develop the currently missing comprehensive legal vehicle. Such a law is strongly preferred over ad hoc and fragmented changes of existing legislation and regulations.

Four pillars should guide the development of energy policy content:

- Bring growth in energy consumption well below the target rate of economic growth.
- Use the country's energy resources on an economically and environmentally sound basis.
- Safeguard the environment from the intensive use of coal.

- Make the energy system more robust so it can withstand supply disruptions.

Urgent policy decisions and strengthened institutions are needed to pursue these pillars and put the energy sector on a sustainable path. Broad energy sector reforms and greater reliance on advanced energy technologies are paramount if sustainability is to be achieved. Stricter standards must be mandated as quickly as practicable to embody energy efficiencies in the new stock of industrial equipment, commercial and residential buildings, and vehicles. Meanwhile, existing standards must be enforced, and retrofit potential must be acted on. China cannot afford delay. The window of opportunity for effective action is closing because of the rapid growth of the stock of energy-consuming goods: industrial equipment, power stations, buildings, vehicles, household appliances, and so forth.

The energy law and high level of policy making should be anchored in the following eight foundational building blocks:

- Create a policy that is integrated, coordinated, and comprehensive.
- Focus strongly on efficiency and immediately on achieving a 20 percent reduction in energy intensity during the 11th Five-Year Plan (2006–10) by quickly defining baselines, national priorities, and sectoral plans.
- Strengthen the institutions for sector governance (giving greater power to the Energy Commission, reestablishing the Ministry of Energy, creating the Energy Education and Information Agency, and improving energy data systems).
- Get the pricing and market fundamentals right, and develop an incentive system to promote sustainability.
- Acquire and deploy cutting-edge energy-efficiency and clean-coal technologies, leapfrogging, when possible, over intermediate technologies, in part by the use of the Clean Development Mechanism (CDM).
- Broaden international cooperation to ease integration into the global market.
- Mitigate potentially adverse social impacts sensitively with targeted measures, including use of emissions reduction revenues

to spur greener and more sustainable development in affected poor areas.
- Mobilize China's civil society and social organizations in pursuit of agreed national policy objectives.

A disconnect exists between the guiding principles adopted and implementation strategies in use. The strategies tend still to be based on command and control and lack the incentives needed to achieve the government's policy objectives. This problem urgently needs to be addressed. Sustainability requires prompt measures but also persistence of effort, with broad diffusion of information and increased public involvement to ensure effective implementation. Increasing public involvement calls for cross-sectoral policy coordination, significant delegation to provinces, and greater involvement of civil society to make energy efficiency a national priority.

At the Threshold of Change toward Sustainability in Energy Structure and Policy

Present energy trends are unsustainable. A sustainable energy system must meet the needs of the present without compromising the ability of future generations to meet their needs. China's energy sector and foreseeable trends in its development plainly do not meet the criteria for sustainability. The intensity of energy use is much higher than in the advanced industrial countries, the energy/gross domestic product (GDP) elasticity has significantly increased during the 10th Five-Year Plan (2001–05), the prospective volume growth is by any standard enormous, the energy mix is dominated by "dirty" coal, oil imports are rising uncontrollably, and the technologies of energy conversion and use are conventional and inefficient. The environmental consequences of these factors are unacceptable in relation to existing Chinese standards. The insecurity implications are threatening.

Present energy policies, institutions, and laws cannot achieve sustainability. Sustainability requires effective policy directions, adequate organizational structures, sound modern legal instruments, and strong economic incentives for efficiency and environmental investments. Established energy policies under which the trends of the 10th Five-Year Plan have developed are clearly deficient and will not lead to sustainability, even though they have achieved much in the way of building a large energy sector, which, with remarkably few failures, has until recently met the volume needs of a rapidly growing economy and a changing society.

Similarly, the existing energy institutions of government are not well adapted to effective management of such a huge sector and its momentous growth. Energy responsibilities, while nominally focused in the National Development and Reform Commission (NDRC), are in effect dispersed among half a dozen departments.[1] These institutions are understaffed and underfunded (box 7.1). Enforcement of laws and regulations is particularly weak.

Regarding the legal instruments for implementing policy, there is no overarching energy law. Deficiencies are reported in regard to regula-

---

Box 7.1

Effective Energy Institutions Involve Significant Staffing and Budgets

In Canada, at the federal level alone, more than 200 people work on energy policies and programs at the Department of Natural Resources, and about 300 are employed by the regulatory agency, the National Energy Board. The combined personnel budgets are about US$64 million annually.

France's electricity regulatory commission had a personnel budget for 2005 of about 120 people, and its operations cost about €30 million annually.

In Kazakhstan, the Agency for the Regulation of Natural Monopolies and Protection of Competition has a staff of 185 to deal with an open access electricity market of about 70 terawatt-hours per year.

The U.S. Department of Energy's responsibilities are diverse and include nuclear weapons. According to the mission statement of the department, its "energy strategic goal" is "to protect our national and economic security by promoting a diverse supply and delivery of reliable, affordable, and environmentally sound energy." The amount budgeted for this purpose in fiscal year 2005 was about US$2.8 billion.

The California Public Utilities Commission, the largest state regulatory body in the United States, is responsible mainly for regulating the gas and electricity industries within the state. It employs about 870 people, and its operating budget in 2005/06 was about US$109 million.

Source: Authors.

tions and norms for the sector. Focused programs to provide strong support, including financial incentives for energy conservation, efficiency standards, and renewable energy development, are still at an embryonic stage. Clearly the policy institutions—organizational and legal—must be strengthened if China is to achieve sustainable energy.

The government of China has recognized the need for change. During the past two years the government has initiated major efforts to increase awareness of the critical need for energy efficiency and to transform society accordingly. Policy makers at the highest level have announced that fundamental changes are needed in the context of a new, comprehensive policy framework that would ensure both long-term energy security and sustainable development.

The four areas for emphasis outlined by the government are sound:

- Focus on energy efficiency and resource savings.
- Encourage further reliance on competition and market forces.
- Enforce the country's laws and regulations.
- Make energy, including efficiency, a national priority by creating a national Energy Commission under the auspices of the State Council.

These four areas provide a viable foundation for a new energy policy.

There is, however, a disconnect between, on the one hand, these recognized policy needs and sound intentions and, on the other hand, their implementation strategies, which still rely on command and control and lack the flexibility needed to unleash creative and locally adapted solutions at lower administrative levels. Also lacking is a sense of urgency and of the need for innovation. Pursuit of a less energy-intensive development path than was followed by other industrial countries is a daunting challenge but not one that is beyond China's capabilities. It requires innovative thinking (no country has yet had the opportunity and political will to embark on such an undertaking), development of a comprehensive policy framework, and a coherent implementation strategy. A low-energy-intensive path will require rethinking and adjusting the economic development strategy, giving consideration to emphasizing less energy-intensive, high value added economic sectors. Such a path must be based on broad public consultation in order to gain the understanding and support of the whole of civil society. Public consultation is particularly important because fundamental choices about ways of life will be needed to achieve sustainability. The urgency arises because the window of opportunity to make fundamental changes is quickly being closed by the huge energy-inefficient additions that are being made to the capital stock of thermal power stations, oil refineries, and commercial and residential buildings and by the tendencies toward suburban living and mass motorization that seem destined to occur as Chinese society follows the trends in other industrial countries.

Policy at the threshold of structural change must embody extraordinary efforts to underpin economic growth with less energy consumption and to steer the energy sector toward a more efficient and therefore less energy-intensive path. This new policy paradigm must be supported by new policy institutions, instruments, and implementation measures. Policy change will have to be brought about without compromising either the continued strong economic growth needed to support rising living standards for China's people or the environmental progress needed to improve quality of life and save the country's limited resources of air, water, agricultural land, forests, and uncongested space. Sustainability must become the hallmark of all of China's energy endeavors and must provide a sharp focus for implementation of the country's laws and regulations.

Characteristics of a Comprehensive Policy for Energy Sustainability

Several factors are needed for an effective energy sustainability policy. First, policy makers, driven by a strong sense of urgency, must make the fundamentally important commitment to strong, effective, and timely decision making. This commitment will provide the essential impetus for these policy implementation steps. Former U.S. Secretary of State Henry Kissinger, having first cited Sun Tzu, recently observed, "China seeks its objectives by careful study, patience, and the accumulation of nuances" (Kissinger 2005, A19). Such traits are admirable, but it is important that they not stand in the way of urgently needed energy policy review, decision making, and implementation actions. Second, sustainability requires long-term vision, careful policy planning, and well-designed enabling policy measures. This effort must go beyond the first, valuable step of targeting a 20 percent reduction in energy intensity during the course of the 11th Five-Year Plan. Creating an integrated, comprehensive, coordinated, and enforceable new energy policy and developing a coherent strategy to implement it should be major priorities for the first years of the 11th Five-Year Plan:

- Energy and environmental policies must be integrated. Possibly the biggest hurdle to achieving a sustainable energy future for China is that environmental externalities are not presently integrated into energy policy, planning, and investment decisions. Full integration of externalities poses huge challenges, but discounting them in the decision-making process is a sure path to unsustainable development. Their importance is dramatically highlighted by the preliminary finding by respected researchers that Guangdong's GDP would be 27 percent lower if measured on a green basis (*People's Daily* 2005).[2]

- Sectoral policies in such varied areas as agriculture, transportation, urban development, foreign and provincial trade, forestry, housing, and mineral resource development must integrate the objectives of energy policy.
- Energy policy must address the needs of all social sectors—rural as well as urban, lower- as well as medium- and higher-income populations, and the western regions as well as the rapidly urbanizing and industrializing eastern provinces.

Third, energy policy must be comprehensive. A comprehensive policy requires addressing the energy needs of all social and productive sectors and securing adequate supplies through all the links of the different fuel chains.

Fourth, there must be horizontal and vertical coordination. In a country of China's size and diversity, an energy policy must allow for flexibility in implementation while upholding vital national choices and standards. The following are critical:

- Central government ministries and agencies need to build consensus on the policy framework and coordinate their activities for its implementation. For example, urban development and transportation policies must be closely coordinated with energy policy because of the strong influences that they have on the quantities and types of energy that will need to be supplied. At present, the state organizations often work in isolation and sometimes send conflicting messages to lower government levels (that is, there is a need for horizontal coordination).
- The provinces and municipalities clearly have important roles to play and should be given more responsibility for charting courses of action and for implementing policy. At present, the provinces and even lower administrative levels are in many cases pursuing short-term local benefits—sometimes driven by a distorted fiscal system—that are detrimental to the national interest and work against the optimal use of resources and energy efficiency.[3] There is a need for vertical coordination

to secure the national interest. Sometimes such coordination will require that strictly local interests should be submerged to take advantage of the synergies and cooperation opportunities between the diverse regions of the country.
- The lower government levels should also be subject to greater accountability. In this connection, like the national institutions, they might need more resources to carry out the detailed studies and analyses that are required to develop sound policies that can be effectively implemented.

Within the allowed regional flexibility, there must be proper enforcement of policies, norms, and implementation measures. Enforcement ensures that national objectives for the energy sector are achieved effectively. Recent history indicates that major choices made at the national level could be undermined by differences between central agencies or weak commitment at lower government levels. Measures for effective enforcement should be worked out and agreed on by all interested parties as part of the new policy.

Finally, energy sustainability is a long-term undertaking, and policy continuity must be ensured, both by political commitment and by the vehicle—namely a comprehensive new law—used to implement it.

Four Guiding Pillars for Sustainability Policy

The four interrelated themes, or pillars, that should guide development of the content of a policy for sustainability—less energy-intensive economic growth, better use of domestic resources, safeguarding of the environment, and enhancement of supply security—can be summarized as follows:

- *Energy consumption growth must be lower than economic growth, to the greatest extent possible.* Achievement of the government's 2020 economic growth target could jeopardize the energy sustainability objective if the energy/GDP elasticity stabilizes at the level reached during the 10th Five-Year Plan or increases beyond it (that is, an elasticity of 1.0 or more). The objective of less energy-intensive economic growth is fundamental: without it, the objectives of ensuring supply security and safeguarding the environment cannot be achieved.
- *Better use must be made of the country's sizable energy resources.* Aside from coal, which will continue to dominate primary energy supply for

the foreseeable future, China's landmass, including its offshore areas, undoubtedly contains further petroleum resources. Some are underused, and others are as yet undiscovered.[4] Greater use of domestic resources can help address insecurity concerns associated with rising oil imports.

- *The environment must be safeguarded from the adverse effects of energy production, conversion, and consumption at the national and local levels.* Environmental degradation caused by the present energy system, particularly by coal use, is at an unacceptable level. It already has resulted in severely adverse effects on human health and well-being, mainly from the presence of particulates in the atmosphere. It is damaging agricultural land, forests, and buildings as a result of acid rain deposition. Strong promotion of energy efficiency and greater use of renewable sources are keys to safeguarding the environment. They could also contribute to a more secure energy supply and mitigate energy price volatility associated with fossil fuels.
- *The energy system must be better prepared to withstand supply disruptions, however caused, and vital energy installations should be made more secure.* China's inevitable and accelerating integration into the global energy economy creates the opportunity to diversify the energy balance and build into it more renewable energy and clean natural gas. At the same time, concerns about security of supply, particularly of oil imports, and about vulnerabilities of critical energy infrastructures such as liquefied natural gas import terminals must be addressed.

These pillars should be used to help assess the value and effectiveness of laws and regulations, programs, and other measures that would provide the eventual content of a comprehensive policy designed to achieve energy sustainability for China. They should also be applied as criteria to examine and evaluate existing energy laws and regulations, leading to their retention, amendment, or abandonment depending on whether their results are consistent with these pillars.

Building Blocks to Put the Energy Sector on a Sustainable Path

Figure 7.1 (see p. 158) sets out the principal features of an organizational structure and relationships to design and implement a coordinated energy policy for China. The central coordinating role could be performed by the Energy Commission. The figure shows the importance of policy and analytical inputs by existing departments, in particular NDRC. The

Figure 7.1 Designing and Implementing a Coordinated Energy Policy

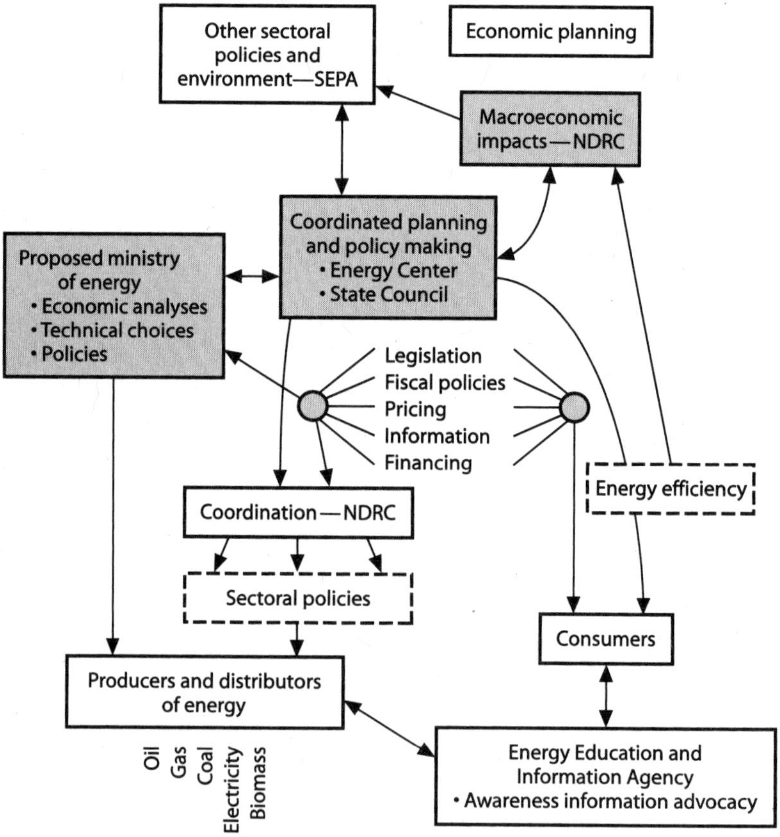

Source: Adapted from Ministère de la Coopération et du Développement 1988.
SEPA = State Environmental Protection Agency.

inputs could be obtained if the Energy Commission makes use of interdepartmental task forces to carry out the studies needed to develop sound policies. The links for analysis and coordination with other sectoral policies are highlighted: they relate to policies such as urban development and transportation, which will be two of the most important drivers of energy consumption in the coming years. The figure emphasizes that consumers should be brought into the policy-making process, with special reference to efficiency considerations. This objective could be accomplished through wide consultation and representation in regulatory and policy hearings. Decision makers need also to consider the creation of an energy ministry to strengthen policy making and to oversee policy implementation.

The energy law under consideration provides the opportunity to embody a coordinated energy policy in a comprehensive, permanent legal vehicle to effect the needed changes and to endow the country with an integrated, coordinated, and enforceable energy policy. Use of such a vehicle is strongly preferred over ad hoc solutions involving existing legislative and administrative authorities.

The following recommendations for a building block approach to energy sustainability, framed around the concept of a new energy law and postulating the Energy Commission as the central policy coordinating institution, are presented for consideration by Chinese authorities. The recommendations relate only to foundational considerations at the first, highest level of policy making. Second-level policy measures, such as oil stockpiling as a means to address energy supply insecurity or particular technologies (for example, integrated coal gasification and power generation or promotion of more efficient cars and trucks) to achieve more efficient and environmentally friendly energy conversion, are recommended or suggested in the report. They will need further assessment and evaluation during the preparation of a first-level energy law.

Building Block 1: Develop an Integrated Policy within a Sound Legal Framework

A comprehensive, coordinated policy directed at energy sustainability should be expressed first in a public commitment made by the highest level of government and then embodied in the energy law under preparation. The public commitment would underscore the government's determination to effect change. It would recognize the long-term nature of the project, and it would stress the importance that will be placed on effective implementation involving broad public consultation, close cross-sectoral coordination with other policies of government (urbanization, agriculture, transportation, and international trade), and a significant degree of delegation to the provinces and lower-level administrative entities within an appropriate framework determined by national laws and regulations. A national Energy Education and Information Agency (EEIA) should be considered. It should have a clear mandate to coordinate all activities designed to make energy efficiency and sustainability a national priority that is supported by all stakeholders, with special focus on civil society (refer to building block 8).

Development of the energy law does not have to wait for the completion of comprehensive programs of studies. What is needed is a framework law that will clearly define the central government's purpose in developing and implementing a comprehensive policy for energy sustainability (see box 7.2 for an example). It should empower the existing institutions and create new ones to complete the needed organizational framework (agencies with clear responsibilities and adequate resources) and give force to the policy that will put the energy sector on a sustainable track and meet the daunting challenges of becoming a resource-conscious society. When technical studies have been completed, consultations carried out, and specific policy implementation decisions made, complementary regulations can be made under the authority of the law, providing for the implementation steps and schedule and ensuring that adequate funding is available for the needed programs.

Building Block 2: Focus Strongly on Efficiency and Sustainability and Consult the Public on These Issues

The purpose of this task is to ensure that efficiency and sustainability are maintained as national priorities. All affected groups—government officials, energy producers, industrial and commercial consumers, and private citizens—must endorse the practicability of what is proposed and accept the behavioral changes required. These groups must be informed and consulted by means of a strong, effective national communications program, developed and implemented by the government, that stresses the strategic and vital nature of energy efficiency in underpinning the concept of China as a resource-conscious society. The communications program could generate valuable feedback from informed audiences regarding novel approaches to achieving the sustainability objective. The

---

Box 7.2

The U.S. Energy Policy Act of 2005

The U.S. Energy Policy Act is a recent example of another industrial country's approach. This legislation builds on a different foundation than China has and reflects a different political culture. However, its scope is impressively broad. There are titles (chapters) dealing with energy efficiency; renewable energy (biomass, ethanol, geothermal, and hydroelectricity); oil and gas; coal; nuclear energy; research and development (buildings, building standards, distributed power, electric-battery vehicles, microgeneration, renewables, and other aspects of energy efficiency); climate change; and incentives for innovative technologies.

Source: Authors.

recommendation under building block 1 to establish an EEIA reflects this expectation.

An immediate priority is to reduce energy intensity by 20 percent by 2010. A first efficiency objective should be vigorous actions to achieve this already targeted reduction in energy intensity. The challenge is daunting: according to 2005 energy and GDP data, over the next five years, China will need to reduce its energy consumption by about 600 million tons of coal equivalent more than in the business-as-usual scenario.[5] This critical objective is nevertheless reachable, but only if drastic and carefully targeted measures are devised and rapidly implemented, because the potential to make easy efficiency gains is shrinking. Achieving such an ambitious program will require an agreement on baselines and monitoring programs, a focus on highly intensive sectors and energy consumption of state-owned enterprises (SOEs), and the involvement of the whole society.

First, officials will have to agree as soon as possible on baselines for measuring current energy intensity—best expressed in terms of grams of coal equivalent per yuan of net output—in aggregate and by sector and on methods for closely monitoring the outcomes of the program. Second, they must establish target reductions in intensity by sector on the basis of sound technical evaluation of the scope for efficiencies. The particular focus should be on highly energy-intensive sectors (ferrous metals, chemical products, and petroleum processing) and the activities of SOEs. Third, they will need to rank the intensity-reduction possibilities in order of economic merit on the basis of cost-benefit analysis. Energy management services companies and policy banks, working in conjunction with business owners, could play an important role by financing economically viable projects. The government may have to provide some financial assistance through concessionary loans through its policy banks or through targeted subsidies. The fourth step is an action program designed to achieve at least the 20 percent reduction by the close of 2010 (see box 7.3 on p. 162). The program would target the sectors, industries, and types of business that offer the best opportunities for efficiency gains. Broadly speaking, three tiers of energy-consuming businesses should be targeted:

- First, SOEs, which typically have not had to face the profitability pressures of listed or private enterprises and tend to use or convert energy wastefully. The SOEs should be subject to command-and-control measures with participation of the energy management companies.
- Second, other businesses where significant energy savings can be achieved by new investments and by changed operating procedures

> **Box 7.3**
>
> **Key Elements of a Program for a 20 Percent Improvement in Energy Efficiency**
>
> A program for 20 percent improvement in energy efficiency requires a number of components:
>
> - Leadership at the highest level with associated public understanding and endorsement
> - Demonstrations of projects and technologies that can produce results quickly
> - Identification and publicizing of national and international best practices in a wide range of uses
> - Challenges issued to consumers at all levels to match best national and international practices
> - Energy-efficiency retrofits for buildings and other energy-using installations
> - Incentives applied selectively to reduce the cost of efficient capital inputs
> - Policies that ensure full-cost pricing of energy inputs to further discourage inefficient uses
> - Tracking of progress, publicity for successes, and attention drawn to failures
>
> *Source:* Authors.

that are fully justified at current energy price levels. There can be an important role here for the government's policy banks in loaning funds beyond the self-financing capability of the enterprises, possibly at concessionary rates.

- Third, another category of businesses where significant energy savings can be achieved, but only by investments that are currently economically marginal unless environmental costs are internalized. In such cases, special measures may be justified to bring efficiency investments in existing and new equipment within the economic threshold. These measures might include incentives such as capital grants; exemption from or reduction in value added tax on inputs or sales; and, where selling prices are still regulated, preapproval to flow through to selling prices the costs of efficiency-increasing investments that may eventually be permitted.

All these program steps need to be taken and the groups of enterprises identified by mid 2007 if this ambitious program is to succeed by the end of the 11th Five-Year Plan (2010).

Building Block 3: Strengthen the Institutions of Energy Policy

The government's commitment to institutional strengthening has been reflected in the establishment of the Energy Commission and the State Electricity Regulatory Commission (SERC) and in the progress made toward creating a modern regulatory structure for the downstream natural gas industry during the 10th Five-Year Plan. The purpose of this building block is to take the process to its logical conclusion. The role of the Energy Commission and its powers should be defined. Consideration should be given to granting it the lead responsibility in advising policy makers on a coherent policy framework with clear objectives, in preparing the energy law that is now under consideration, and in devising the information and consultation process in building block 2. Drawing also on the expertise of other departments and agencies through the creation of task forces, the Energy Commission could be responsible for a program of studies in areas that are vital to sustainable development, such as the following:

- Pricing and taxation issues, with a view to incorporating external environmental costs in energy pricing and to modernizing energy taxation
- Evaluation and recommendation of technology choices, with particular attention to clean coal and to novel approaches to urbanization and personal transportation
- Enhanced international cooperation to support national objectives such as technical leapfrogging, other technology transfer, and improved supply security.

With the Energy Commission mandated and active, the next step is a comprehensive review of the current government organization for energy and identification of changes needed to meet the challenges of sustainability. In this connection, consideration should be given to these tasks:

- Reestablish a ministry of energy (see box 7.4 on p. 164).
- Properly empower the SERC and clarify its responsibilities—in particular its authority to access the accounts and financial statements of regulated companies and its responsibilities for pricing in the power subsector. Clarification of whether the SERC should be responsible for policy and for policy implementation or whether the policy function should be retained by a government agency is especially needed.
- Create a regulatory body (or an autonomous department within the SERC) for the downstream gas industry.

> **Box 7.4**
>
> **Creation of a Strong Energy Ministry Reflects International Practice**
>
> International practice is to provide for a strong energy department with policy-making and implementation mandates and an appropriate budget. Since the global oil supply emergency of 1973 and 1974, the governments of most industrial countries have established energy departments or ministries to do this job. For example, the United States has the Federal Department of Energy, created in the late 1970s. In France and in the United Kingdom, energy is the responsibility of a director-general or junior minister within a superdepartment of industry. In many other countries, energy is the responsibility of a self-contained sector within a department of natural resources or, perhaps, within a ministry of energy and mines. In almost all instances, energy matters are a cabinet-level responsibility.
>
> *Source:* Authors.

- Create other agencies parallel to or within the EEIA, especially to promote energy efficiency and technological innovation to achieve sustainability.

Institutional strengthening will also require that existing deficiencies in human and financial resources be addressed in a timely manner. Doing so will require larger budgetary allocations for energy policies and programs within the central government. Attention also needs to be given to improving the technical capability of regulatory institutions, their practices in terms of fairness and transparency, and issues relating to their authority over regulated utilities (for example, the matter of requiring periodic financial and operational reporting in accordance with mandatory uniform accounting standards).[6]

A particular institutional consideration is improving historical energy data and data flows. The situation in regard to both historical energy data and current data flows is unsatisfactory. The data available at present appear to be unreliable and have been subject to revision long after the fact. In terms of coverage, integration, and timeliness, China's energy data sources are inferior to those of other industrial and most industrializing countries. This deficiency, which is elaborated in appendix D, handicaps the analysis needed for sound policy development and, unless the situation is improved, will make it difficult to carry out close and effective monitoring of the changing energy picture—including the effects of policies for sustainability—to determine whether and what midcourse changes may be needed. Data deficiencies are particularly troublesome in relation to the objective of the 20 percent reduction in energy intensity by the end of the 10th Five-Year Plan. Developing a strong statistical system that can produce reliable energy balances and other information on a timely basis is recommended as a prerequisite for sound planning and policy making.

Building Block 4: Get the Fundamentals of the Sector Right

A policy for energy sustainability needs to harness the power of markets and market pricing so that correct signals go out to energy producers, importers, exporters, consumers, and investors to bring about needed changes in patterns of production, consumption, trade, and investment, to meet energy requirements in the most efficient way. At the same time, targeted elements of command and control (consumption taxes, and efficiency standards) will be needed to complement or correct the working of markets. In the first place, the unfinished business of price reform needs to be completed (a) by removing the remaining controls on energy commodity prices (power, oil products, and natural gas), thereby allowing markets to form prices, while monitoring for abuse of monopolistic or dominant positions and (b) by completing the reform of natural monopolies in pipes and wires, thus providing for modern regulation of the prices and terms of access for their services. In the second place, the transition to competitive markets for energy commodities must be completed by allowing and encouraging more buyers and sellers. In the third place, external costs, mainly environmental, must be incorporated in prices as soon as practicable and at a rate that is related to the capacity of the energy sector to adjust. In every energy industry, the guiding principle for planning, investment, and regulation must be cost minimization (sometimes termed *least-cost planning and investment*). This principle is best achieved through market mechanisms, except where natural monopoly exists or producers have market power. Development of building block 4 would be guided by the results of the program of studies called for in building block 3.

Building Block 5: Deploy Cutting-Edge Technologies and Enhance Technology Transfer, Development, and Implementation

The basic purpose of this task is to identify, assess, and obtain best-in-class technologies applicable across the whole energy sector and secure their widest and most rapid possible application in China. New technologies are the essential means to sever energy growth from economic growth without impairing living standards, to improve security by reducing import needs, and to deal with energy-related environmental degradation. Clean, climate-friendly coal technologies are a particular focus for action, as is the issue of personal mobility. These technologies may require larger upfront capital spending than traditional ones. They should be evaluated on the basis of cost-benefit analyses over the life cycle of the assets and environmental externalities. Trading of low-cost carbon emission reductions under the CDM of the Kyoto Protocol provides a good opportunity to fund access to advanced energy efficiency and clean-coal technologies. Development of a comprehensive strategy to deploy cutting-edge technologies should be based on comprehensive technical assessments and in-depth studies, including the economic and fiscal incentives for their deployment. In some cases, these technologies are not yet applied commercially in the industrial countries, for reasons that include concerns about stranded costs linked to past policies and markets, lack of opportunity because of low growth, and the presence of vested interests.

Building Block 6: Broaden International Cooperation

The purpose of this building block is to help achieve China's national interests in energy supply security and in all areas of sustainable energy development, but particularly in technology transfer. International experience over many years clearly shows that multilateral and bilateral cooperation can help secure supply from international markets and can minimize negative impacts in times of crisis. International cooperation is one means to secure the most advanced technologies. There are already cooperative ventures with the European Union, Japan, and the United States in zero-emissions coal technology.

Some of the relevant technologies are already being implemented in industrial countries (for example, unconventional natural gas, including coal-bed methane). However, there are undoubtedly other technologies developed in such countries that may not quickly achieve acceptance because of social reasons (populations habituated to large personal vehicles and low use of public transportation); because of established infrastructure that militates against fundamental fuel-use changes (such as hydrogen fuel cell vehicles); or because of market dominance of a particular technology (for example, electric rather than gas- or renewable energy–based air conditioning). In these circumstances, foreign developers of such technologies might be particularly attracted by the

> **Box 7.5**
>
> ### China and the International Energy Agency
>
> Although China is not a member of the International Energy Agency (IEA), China and the IEA have mutually and fruitfully pursued closer relations in the past five years. Areas of cooperation have included coal industry investment, electric power reform, emergency response training, energy modeling, energy relations in northeast Asia, energy-efficiency standards, and labeling. In parallel with the development of bilateral energy relations, the United Nations, IEA, and, regionally, Asia Pacific Economic Cooperation and possibly Association of Southeast Asian Nations offer valuable opportunities for multilateral cooperation and coordination. New possibilities for technology transfer may also be opened up by the recently created Asia-Pacific Clean Development and Climate Change Partnership, which is intended to complement the Kyoto Protocol and the United Nations Climate Change activities and has been strongly welcomed by the government.

possibilities of the Chinese market. There is therefore scope for mutual benefits. Foreign technology developers would be able to demonstrate their products, probably on a scale not possible elsewhere, leading to local manufacturing of the products. Chinese experts and enterprises would be exposed to these technologies and could gain experience with them by partnering in joint ventures or even by leading development. China's energy economy stands to gain from new efficiencies in production, conversion, and use. Creative technical solutions are needed to enable China to access and deploy technologies that would contribute to important reductions in local pollution and in greenhouse gas emissions. In chapter 4, attention was drawn to the opportunities presented by advanced energy technologies to safeguard the environment and advance the cause of sustainability and the importance of seeing to it that China's access to these technologies is not impeded by any failure of international cooperation (see also box 7.5).

Building Block 7: Mitigate the Impacts of Energy Reforms on Vulnerable Groups Sensitively and with Targeted Measures

Achieving energy sustainability will entail sweeping policy changes and reforms. Some of these changes will have specific social and regional impacts. Despite the overall benefits of those changes, some of them may adversely affect particular social groups (such as low-income consumers of gas and electricity) or particular localities (such as cities and regions where energy-inefficient facilities such as small oil refineries and coal mines are subject to closure or rationalization). The objective of this building block is to anticipate, inventory, and address in advance specific social and regional effects of policies for sustainability. In general, it will likely be most efficient and economical to address these issues with targeted measures such as direct income assistance or lifeline rates for low-income consumers rather than to delay or distort sensible comprehensive reforms. This is an area where elements of command and control may be needed to complement or correct the free working of markets. Innovative approaches should be explored, such as selling emissions reductions to generate revenues that can be used to address social concerns and create alternative development strategies before closing small and polluting power plants at the county level.

Building Block 8: Mobilize China's Civil Society and Social Organizations in Pursuit of Agreed National Policy Objectives

Civil society organizations (CSOs, called *social organizations* in China) are nongovernmental organizations that can play a significant role in the delivery of services and implementation of development programs, especially where the government's presence may be weak and where the CSOs' expertise complements the government's actions.[7] By way of an example of their importance on a global scale, the projected CSO involvement in World Bank–funded projects has risen steadily over the past decade, from 21 percent of all projects in fiscal year 1990 to an estimated 72 percent in fiscal year 2003.

In terms of a policy for energy sustainability for China, CSO involvement can contribute positively in the following ways:

- It can give a voice to stakeholders, perhaps to poor and marginalized populations in rural and western parts of the country, thus ensuring that their views are factored into energy policy and program decisions.
- It can valuably promote public consensus and local ownership for energy policy reform measures by building common ground for understanding and encouraging public-private cooperation.
- It can bring in innovative ideas and solutions, as well as participatory approaches to solving local problems.

- It can strengthen and leverage development programs by providing local knowledge, targeting assistance, and generating social capital at the community level.
- It can provide professional expertise and increased capacity for effective service delivery, especially in environments with weak public sector capacity or in postconflict contexts.

For all these reasons, the potential for involving CSOs should be carefully assessed, with a view to shaping policies for sustainability and achieving effective implementation attuned to local needs.

The Broad Challenge of the Building Block Approach
Each of the eight building blocks represents a major undertaking or activity. Each will require factual studies, consultation with experts, analysis of options, and recommendations to and decisions by policy makers. The development of an energy law will involve much work by legal staff members from many parts of the government, decisions by the State Council, and proposals to the People's Congress. Public consultation will have to be based on carefully prepared and targeted messages and, to be effective, will involve identifying, briefing, and interacting with key public groups. Institutional strengthening will probably require significant reorganization and rejuvenation of the central government's capability to make and implement energy policy. The ground is already well prepared for getting the fundamentals of pricing and markets right, but implementation will take political commitment and will require changing mindsets and the ingrained control culture. Enhancing technology transfer and development will present some major scientific, engineering, and social challenges and will probably involve some significant government seed money to initiate the development and implementation process. Broadening international energy cooperation means mobilizing diplomatic and commercial resources and expertise to achieve a variety of goals in a well-coordinated campaign. The sensitive use of targeted measures is going to be necessary to effect smooth implementation of policies for sustainability where social and local impacts may be significant. Finally, the possible mobilization of CSOs to help shape and implement policies for sustainability will require identifying target groups, inventorying tasks on which their assistance would be appropriate, and continuing project liaison and monitoring. The most profound challenge will be to carry out these tasks within the closing window of opportunity.

The Sequence of Steps to Sustainability

The first step is for the government to vigorously reiterate its commitment to an integrated energy policy set in a sound legal framework that is directed at the overarching goal of sustainability. This commitment should be expressed in a policy statement by the highest level of government. It should include references to consultation, to the key role the provinces and lower-level administrative entities are expected to play, and to the intention to consult widely.

In the short term (first 12 months),

- The Energy Commission should be staffed, resourced, and given these mandates:
  —Advise on a new, comprehensive energy law.
  —Consult the public and report back.
  —Coordinate a profound improvement in the historical and current energy data in terms of coverage, accuracy, and timeliness.
  —Supervise a program of energy studies carried out by interdepartmental task forces and staffed by the nation's leading energy experts from government, industry, and academia.
  —Secure close cooperation and coordination with other government departments and the policies for which they are responsible, with particular reference to policies on foreign trade, transportation, and urban development.
- Existing price controls on energy commodities should be reviewed and removed as soon as possible to prevent giving distorted signals to consumers and to encourage the responsible use of energy.

In the short term, but extending into the balance of the 11th Five-Year Plan if necessary,

- The new framework energy law should be enacted so that it can provide a comprehensive basis for action, in terms of making regulations and authorizing spending, in all of the subjects required to implement a sustainable energy policy.[8]
- An aggressive program to reach the target of a 20 percent reduction in energy intensity by 2010 should be devised and launched. These tasks will be involved:
  —Agree on baselines and on monitoring of the outcomes (soonest).
  —Establish target reductions by sector, focusing on highly energy-intensive ones and targeting the large SOEs.

—Rank those reductions by industry and business.
—Initiate a program of incentives for investments and changed operating practices to achieve the target reductions, with government tax and other assistance to be available where justified.

All these steps need to be taken by mid 2007 if this ambitious program is to succeed by the end of the 11th Five-Year Plan (2010).

In the medium term, beyond 12 months, but with some work to be initiated as soon as possible,

- The role and responsibilities of the SERC should be clarified and the SERC empowered to work toward having a functioning electricity market in which transaction and investment behaviors are progressively modified by the incorporation of environmental externalities in pricing and in investment decisions.
- A gas regulatory commission should be created (possibly within a department but with ample autonomy from the SERC) with responsibilities for the gas market downstream of the gas purification plants similar to those the SERC has for the electricity market. As a result, natural monopolies in the gas and electricity industries (the transmission and distribution functions) would be subject to modern regulation.
- The opening of commodity markets for electricity, gas, and oil products should be completed, competition should progressively increase, and authorities should monitor (rather than regulate) prices to check for possible abuse of dominant positions. As a result, energy commodity markets would be regulated by competition rather than by the government, with consequent efficiencies in the use of all resources and better balancing of supply and demand in all circumstances.
- A program of technology transfer, development, and rapid implementation should be launched. As a result, best-in-class technologies from around the world would be built into China's energy economy with consequent efficiency and environmental benefits.
- International cooperation, bilateral and multilateral, should be intensified.
- Improvement in the flexibility and security of energy supply should be a continuing, overarching objective helping to guide decision making in such varied areas as international cooperation, energy technology, and energy commodity sourcing.
- The government should respond sensitively to social and local issues stemming from energy reform.

- China's CSOs should be harnessed to the task of consulting, informing, and reacting to proposals for energy policies, programs, and measures and to the task of implementing policies for energy sustainability where they can make an effective contribution, particularly in remote and rural areas.

Throughout the short and medium terms, policy implementation should be characterized by a strong sense of urgency, motivated by the fact that the window of opportunity to achieve profound technical and social changes is closing rapidly.

In the long term, beyond the 11th Five-Year Plan,

- The Energy Commission, using the best research and advice available, and building on the experience of the 11th Five-Year Plan, should draw up a program of continuous long-term improvement in energy and environmental efficiency and sustainability.
- The energy law should be continuously updated as necessary to ensure that its provisions are relevant to the developing energy economy and that they provide all the powers needed to continue along the path of sustainability.
- Exploration and development of energy resources—oil, gas, coal, and renewables—should be exposed to investment and entrepreneurship without restriction as to source, whether public or private, domestic or foreign.[9]
- The government's activity in energy commodity markets should be limited to careful monitoring of the functioning of those markets and intervention only in the case of demonstrated abuse of dominant market positions by sellers or buyers.
- The application of the best technologies available internationally should be mandatory in all major new energy production, conversion, and consumption investments and for vehicles and appliances for sale in the domestic market.

This program of action has these expected outcomes:

- In the short term, the senior-level commitment to a policy of energy sustainability will provide valuable direction for change. The Energy Commission's activity, carried out through task forces in the central government and through consultation with the provinces and public

groups, will help secure previously lacking horizontal and vertical coordination in preparing proposals for consideration by policy makers. The program of studies will provide the basis for a bank of factual information and ideas on which policy makers could draw to develop the energy law. Meanwhile, as a result of administrative actions, progress will resume toward the establishment of functioning markets for energy commodities.
- In the short term and into the term of the 11th Five-Year Plan, the launch of a well-conceived program aimed at a 20 percent reduction in energy intensity by 2010 will provide a rallying point for efficiency efforts. It will be the first core program to lower energy needs per unit of output without compromising economic growth, hence achieving relative improvements in the environment and a relative reduction in import needs. The 20 percent reduction program will establish a template for subsequent efforts—for example, in technology transfer—in terms of initial analysis, targeting, delegation, and provision of technical and financial support for implementation decisions made at a local level.
- In the medium term, valuable efficiencies will be achieved by the better operation of markets in the gas and electricity industries, some of which will contribute also to environmental improvements. There will be, as well, an effect of rapid gains in energy efficiency across all sectors, thereby tending to keep energy growth at a significantly lower rate than economic growth. The relationship between growth in energy consumption and growth in personal income will be weakened by new policies relating to urbanization and mass transportation as well as by higher efficiency standards for buildings, appliances, and vehicles. Maximizing national strengths in a focused program of greater bilateral and multilateral international cooperation should yield benefits in terms of creating more secure imported energy supplies, establishing greater diversification of those supplies, and harnessing to Chinese needs programs of technology transfer and various forms of emissions crediting. Measurable improvements in energy intensity and related reductions in energy elasticity will be reported, resulting in lower-than-otherwise aggregate energy needs, import requirements, and emissions of pollutants.
- In the longer term, China will be equipped with a modern, comprehensive energy law that fully reflects the government's commitment to energy sustainability and that gives confidence to all actors—

investors, consumers, and producers—in the permanence of this policy direction. The energy mix, although dominated by increasingly clean coal, will diversify and include a growing proportion of natural gas, and the share of electricity in final consumption will continue to rise. At the same time, electricity generation will have a growing renewable energy content. Energy markets will function effectively: they will exhibit a reasonable degree of transparency, abuse of dominant positions will be curbed, and natural monopolies will be efficiently regulated. Any adverse impacts of market pricing and other policy elements on particular economic, social, and geographic sectors will be effectively dealt with by targeted programs such as special rate schedules, subventions, and income assistance. Confidence in the effectiveness and independence of China's regulatory institutions and the competitive capability of its energy companies will rise to a point where policy can allow broad access by investors of all kinds to the development of resources in China. The result: these resources will be exploited more intensively and efficiently than they would be by the listed SOEs acting alone. At the same time, the listed SOEs will be integrating more smoothly into the international scene. They will act commercially and bring fresh ideas and new capital. Consequently, they will enjoy improved access to foreign investment opportunities and sharpen their competitive abilities. There will be a well-publicized ongoing program to build into the economy continuous improvement in energy efficiency that will cumulate and advance the gains achieved during the 11th Five-Year Plan. The understanding and application of the best available and most efficient and appropriate technologies will be such that most new investment, whether large-scale (new thermal electric plants) or small-scale (new rural home construction), will incorporate those cutting-edge technologies. Meanwhile the existing stock of energy-using and conversion equipment will—to the greatest extent possible—be progressively modernized, with significant resulting energy savings. Finally, China will be recognized internationally as a proving ground for new, efficient energy technologies and become a net exporter of products that use them.

Table 7.1 (see pp. 176–77) summarizes these points.

Notes

1. Administrations, bureaus, commissions, and ministries that have energy responsibilities include the Ministry of Commerce, Ministry of Finance, Ministry of Foreign Affairs, Ministry of Land and Resources, State Environmental Protection Administration, and Price Bureau.
2. According to a newspaper article, a report published by the Guangzhou Institute of Geochemistry of the Chinese Academy of Sciences and the Guangdong Institute of Eco-Environment and Soil Sciences found that the GDP of Guangdong province would have been 27 percent lower in 2003 if it had been adjusted for environmental costs (*People's Daily* 2005).
3. It is well documented that provinces are reluctant to rely on electricity imported from other parts of China for two main reasons. First, fiscal revenues related to the value added tax are collected at the generation level rather than at the consumption level. This practice provides the wrong incentives to authorities at the subnational level, who prefer to build generation assets in their administrative areas rather than importing what may be cleaner (and even cheaper) electricity. Second, the self-reliance syndrome (which characterized China's development strategies at earlier stages) is still well ingrained, especially in an environment where contract enforcement remains weak.
4. Regarding petroleum resources, in addition to conventional oil and gas, onshore and offshore undiscovered resources may include oil sands and oil shale, tight gas, shale gas, and coal-bed methane.
5. This reduction amount is greater than the 2004 primary energy consumption of every other country of the world, with the exception of Japan, the Russian Federation, and the United States.
6. The extent of information-gathering powers of regulatory authorities is a consistent subtheme in the joint reports of the World Bank and the Institute of Economic System and Management of the former State Council Office for Restructuring the Economic System. See, for example, the English-language version of World Bank and IESM (2001, 195, annex 5, item 21): "information gathering according to a uniform system of accounting." See also World Bank and IESM (2002a, 233, section 3.3): "Gather information, including the authority to compel the provision of information by regulated entities."

Table 7.1 Shaping the Future: The Issues, Guiding Pillars, Building Blocks, and Sequencing Steps to Sustainability

Fundamentals: trends toward an unsustainable future → Conclusion: present trends unsustainable

Five outstanding issues:
- Scale of needs
- Scale of environmental damage
- Import insecurity and electricity reliability
- Policy disconnects
- Policy implementation

Sustainability: challenging but feasible → Conclusion: paradigm shift needed

Four pillars to guide sustainability:
- Create energy growth lower than economy growth
- Safeguard environment
- Harness resources and prepare for a smooth integration in international energy markets
- Prepare to withstand disruptions

Path to sustainability → Conclusion: initiate integrated, novel, and coordinated policies

Start with a policy commitment; then pursue recommendations for eight building blocks to pave way to sustainability:
- Create an integrated policy in a sound legal framework including an energy ministry and education and information agency
- Focus strongly on efficiency and immediately on achieving 20 percent reduction in energy intensity during 11th Five-Year Plan
- Strengthen the institutions for sector governance (that is, define role of Energy Commission and empower SERC) and improve energy data flows
- Get the sector fundamentals—markets and prices—right
- Enhance technology transfer and development, and speed up deployment (especially clean coal)
- Broaden international cooperation (in security and technology)
- Mitigate potentially adverse impacts on vulnerable groups with targeted measures
- Mobilize civil society organizations to pursue agreed national objectives

Sequencing steps

Short-term steps:
- Establish government commitment to integrated energy policy
- Mandate Energy Commission (advise on law, consult public, and supervise studies)
- Review price controls, and substitute monitoring
- Carry out program of studies

Short-term steps and steps into term of 11th Five-Year Plan:
- Implement aggressive program to reduce intensity by 20 percent by 2010:
  —Agree on baselines and monitor progress
  —Establish target reductions
  —Rank those reductions
  —Develop communications strategy
  —Initiate program with incentives (all by end of year 1 of 11th Five-Year Plan)

Table 7.1. Shaping the Future: The Issues, Guiding Pillars, Building Blocks, and Sequencing Steps to Sustainability—Continued

Medium-term steps:
- Empower and clarify responsibilities of SERC
- Create downstream gas regulator (regulate monopoly prices and access)
- Complete opening of commodity markets (markets regulate commodity prices, which increasingly include externalities)
- Launch and rapidly implement program of technology transfer
- Intensify international cooperation, bilateral and multilateral
- Sensitively respond to social and local issues stemming from energy reform

Long-term steps:
- Establish program for continuous efficiency improvement
- Review energy law to ensure continued adequacy and relevance
- Open energy resources to all investors
- Limit government activity in commodity markets to monitoring and correction
- Implement mandatory application of best available energy-efficient technologies in new investments

Anticipated outcomes

Short-term outcomes:
- Responsibility center created
  - Achieve horizontal coordination across central government (task forces)
  - Achieve vertical coordination with provinces and public groups (consultation)
- Progress toward functioning markets resumed
- Bank of facts and ideas created to draw on for lawmaking

Short-term outcomes, into 11th Five-Year Plan:
- 20 percent established as rallying point and core program created to delink growth in energy and economy, improve environment, and enhance security
- Model for subsequent programs (such as technology transfer) created in terms of analysis, targeting, delegation of implementation, and provision of technical and financial support

Medium-term outcomes:
- Market efficiencies achieved in power and gas subsectors
- Rapid gains in energy technology achieved across all sectors that delink growth of energy and economy
- New policies created on urbanization and transportation that delink energy and income growth
- Diversification increased
- Maximum advantages from international cooperation achieved by levering national strengths

Long-term outcomes:
- Energy law reviewed to confirm permanence of sustainability commitment
- Cleaner, more secure energy mix achieved
- Domestic energy resource development enhanced
- Continuous improvement achieved, best available technologies applied to major new investments, and existing plant updated for further efficiency and environmental gains.

Source: Authors.

Finally, see World Bank and IESM (2002b, 260–61, articles 17 and 18): powers "to gather information, including compelling the provision of information from any license holder."
7. The World Bank uses the term *civil society* to refer to the wide array of nongovernmental and not-for-profit organizations that have a presence in public life, expressing the interests and values of their members or others, based on ethical, cultural, political, scientific, religious, or philanthropic considerations. The term *CSOs* therefore refers to a wide of array of organizations: community groups, nongovernmental organizations, labor unions, indigenous groups, charitable organizations, faith-based organizations, professional associations, and foundations. The relevant website is http://web.worldbank.org/WBSITE/EXTERNAL/TOPICS/CSO/0,,contentM DK:20101499~menuPK:244752~pagePK:220503~piPK:220476~theSitePK:228717,00.html.
8. The main headings of the energy law might include (in alphabetical order): coal, including clean coal; consultation mechanisms; conventional energy forms, other than coal; cooperation and coordination with the provinces and large cities; efficiency standards; environmental protection; government organization, including regulatory institutions; international cooperation; international trade; monitoring and public reporting; nuclear energy; prices and price controls; regulation of natural monopolies; renewable forms of energy; research and development; security measures; and state-owned enterprises.
9. The case of hydroelectric resources would have to be subject to special consideration. Because of the unusual characteristics of hydroelectricity, for example, the creation of economic rents that are difficult to assess and collect, there is an argument for reserving its development to SOEs.

References

Kissinger, Henry A. 2005. "China: Containment Won't Work." *Washington Post*, June 13.

Ministère de la Coopération et du Développement. 1988. "Guide de l'Énergie." Ministère de la Coopération et du Développement, Paris.

People's Daily. 2005. "Guangdong Pioneers New GDP Model." July 20. http://english.people.com.cn/200507/20/eng20050720_197250.html.

World Bank and IESM (Institute of Economic System and Management). 2001. *Modernizing China's Oil and Gas Sector—Structure Reform and Regulation.* World Bank: Washington, DC. http://www.worldbank.org.cn/English/content/oil-gas.pdf.

———. 2002a. *China: Economic Regulation of Long-Distance Gas Transmission and Urban Gas Distribution.* Washington, DC: World Bank.

———. 2002b. *China: Regulatory Framework for Long-Distance Gas Transmission and Urban Gas Distribution.* Washington, DC: World Bank.

APPENDIX A

Gross Domestic Product and Energy Consumption in China, 1980–2005

	1980	1981	1982	1983	1984	1985	1986	1987	1988	1989	1990	1991	1992
Part 1: GDP													
GDP (original)													
Nominal (Y billion)	452	486	529	593	717	896	1,020	1,196	1,493	1,691	1,855	2,162	2,664
2000 constant price (Y billion)	1,405	1,479	1,612	1,787	2,059	2,336	2,543	2,837	3,157	3,285	3,411	3,725	4,255
Annual growth rate (%)	7.8	5.2	9.1	10.9	15.2	13.5	8.8	11.6	11.3	4.4	3.8	9.2	14.2
GDP by sector (calculated at nominal price)													
Primary (%)	30.1	31.8	33.3	33.0	32.0	28.4	27.1	26.8	25.7	25.0	27.1	24.5	21.8
Secondary (%)	48.5	46.4	45.0	44.6	43.3	43.1	44.0	43.9	44.1	43.0	41.6	42.1	43.9
Tertiary (%)	21.4	21.8	21.7	22.4	24.7	28.5	28.9	29.3	30.2	32.0	31.3	33.4	34.3
GDP (revised in 2005)													
Nominal (Y billion)	452	486	529	593	717	896	1,020	1,196	1,493	1,691	1,855	2,162	2,664
2000 constant price (Y billion)	1,515	1,595	1,739	1,928	2,221	2,520	2,743	3,060	3,405	3,544	3,680	4,018	4,590
Annual growth rate (%)	7.8	5.2	9.1	10.9	15.2	13.5	8.8	11.6	11.3	4.4	3.8	9.2	14.2
Part 2: Energy consumption													
Energy consumption (original)													
Total consumption (million tce)	603	594	626	660	709	767	809	866	930	969	987	1,038	1,092
Annual growth rate (%)	2.9	−1.4	5.4	5.4	7.4	8.1	5.4	7.2	7.3	4.2	1.8	5.1	5.2
Share of energy consumption													
Coal (%)	72.2	72.7	73.7	74.2	75.3	75.8	75.8	76.2	76.2	76.0	76.2	76.1	75.7
Crude oil (%)	20.8	20.0	18.9	18.1	17.5	17.1	17.2	17.0	17.1	17.1	16.6	17.1	17.5
Natural gas (%)	3.1	2.8	2.6	2.4	2.4	2.2	2.3	2.1	2.1	2.1	2.1	2.0	1.9
Hydroelectric and nuclear power (%)	4.0	4.5	4.9	5.3	4.9	4.9	4.7	4.6	4.7	4.9	5.1	4.8	4.9

	1980	1981	1982	1983	1984	1985	1986	1987	1988	1989	1990	1991	1992
Part 2: Energy consumption (continued)													
Energy consumption (adjusted in 2006)													
Total consumption (million tce)	1,160	1,227	1,312	1,389	1,378	1,322	1,338	1,386	1,432	1,518	1,750	2,032	2,225
Annual growth rate (%)	2.9	−1.4	5.4	5.4	7.4	8.1	5.4	7.2	7.3	4.2	1.8	5.1	5.2
Part 3: Electricity consumption													
Electricity consumption (TWh)	301	310	328	352	378	412	451	499	547	587	623	680	759
Annual growth rate (%)	6.6	3.0	5.9	7.3	7.4	9.0	9.4	10.6	9.7	7.3	6.2	9.2	11.5
Part 4: Intensity (based on 2000 constant price)													
Energy intensity													
Original energy/original GDP (tce/Y 1,000)	0.429	0.402	0.389	0.369	0.344	0.328	0.318	0.305	0.295	0.295	0.289	0.279	0.257
Original energy/revised GDP (tce/Y 1,000)	0.398	0.373	0.360	0.343	0.319	0.304	0.295	0.283	0.273	0.274	0.268	0.258	0.238
Adjusted energy/original GDP (tce/Y 1,000)	0.429	0.402	0.389	0.369	0.344	0.328	0.318	0.305	0.295	0.295	0.289	0.279	0.257
Adjusted energy/revised GDP (tce/Y 1,000)	0.398	0.373	0.360	0.343	0.319	0.304	0.295	0.283	0.273	0.274	0.268	0.258	0.238
Electricity intensity													
Electricity consumption/original GDP (kWh/Y)	0.214	0.209	0.203	0.197	0.184	0.176	0.177	0.176	0.173	0.179	0.183	0.183	0.178
Electricity consumption/revised GDP (kWh/Y)	0.198	0.194	0.189	0.183	0.170	0.163	0.164	0.163	0.161	0.166	0.169	0.169	0.165

(continued)

	1980	1981	1982	1983	1984	1985	1986	1987	1988	1989	1990	1991	1992
Part 5: Elasticity													
Original energy vs. original GDP	0.37	−0.26	0.59	0.50	0.48	0.60	0.62	0.62	0.65	0.96	0.48	0.56	0.37
Original energy vs. revised GDP	0.37	−0.26	0.59	0.50	0.48	0.60	0.62	0.62	0.65	0.96	0.48	0.56	0.37
Adjusted energy vs. original GDP	0.37	−0.26	0.59	0.50	0.48	0.60	0.62	0.62	0.65	0.96	0.48	0.56	0.37
Adjusted energy vs. revised GDP	0.37	−0.26	0.59	0.50	0.48	0.60	0.62	0.62	0.65	0.96	0.48	0.56	0.37
Electricity vs. original GDP	0.85	0.58	0.65	0.67	0.48	0.67	1.07	0.91	0.86	1.65	1.64	1.00	0.81
Electricity vs. revised GDP	0.85	0.58	0.65	0.67	0.48	0.67	1.07	0.91	0.86	1.65	1.64	1.00	0.81

	1993	1994	1995	1996	1997	1998	1999	2000	2001	2002	2003	2004	2005
Part 1: GDP													
GDP (original)													
Nominal (Y billion)	3,463	4,676	5,848	6,788	7,446	7,835	8,207	8,947	9,731	10,479	11,739	13,688	18,232
2000 constant price (Y billion)	4,829	5,441	6,013	6,589	7,171	7,732	8,284	8,947	9,618	10,416	11,385	12,466	13,700
Annual growth rate (%)	13.5	12.6	10.6	9.6	8.8	7.8	7.1	8.0	7.5	8.3	9.3	9.5	9.9
GDP by sector (calculated at nominal price)													
Primary (%)	19.9	20.2	20.5	20.4	19.1	18.6	17.6	16.4	15.8	15.4	14.4	15.2	12.5
Secondary (%)	47.4	47.9	48.8	49.5	50.0	49.3	49.4	50.2	50.1	51.1	52.2	52.9	47.3
Tertiary (%)	32.7	31.9	30.7	30.1	30.9	32.1	33.0	33.4	34.1	33.5	33.4	31.9	40.3
GDP (revised in 2005)													
Nominal (Y billion)	3,533	4,820	6,079	7,118	7,897	8,440	8,968	9,922	10,966	12,033	13,582	15,988	18,232
2000 constant price (Y billion)	5,233	5,918	6,563	7,219	7,891	8,506	9,153	9,922	10,745	11,723	12,895	14,197	15,603
Annual growth rate (%)	14.0	13.1	10.9	10.0	9.3	7.8	7.6	8.4	8.3	9.1	10.0	10.1	9.9

	1993	1994	1995	1996	1997	1998	1999	2000	2001	2002	2003	2004	2005
Part 2: Energy consumption													
Energy consumption (original)													
Total consumption (million tce)	1,160	1,227	1,312	1,389	1,378	1,322	1,301	1,303	1,349	1,482	1,709	1,970	2,225
Annual growth rate (%)	6.2	5.8	6.9	5.9	−0.8	−4.1	−1.6	0.1	3.5	9.9	15.3	15.2	12.9
Share of energy consumption													
Coal (%)	74.7	75.0	74.6	74.7	71.5	69.6	68.0	66.1	65.3	65.6	67.1	—	—
Crude oil (%)	18.2	17.4	17.5	18.0	20.4	21.5	23.2	24.6	24.3	24.0	22.7	—	—
Natural gas (%)	1.9	1.9	1.8	1.8	1.8	2.2	2.2	2.5	2.7	2.6	2.8	—	—
Hydroelectric and nuclear power (%)	5.2	5.7	6.1	5.5	6.3	6.7	6.6	6.8	7.7	7.8	7.4	—	—
Energy consumption (adjusted in 2006)													
Total consumption (million tce)	603	594	626	660	709	767	809	866	930	969	987	1,038	1,092
Annual growth rate (%)	6.2	5.8	6.9	5.9	−0.8	−4.1	1.2	3.5	3.4	6.0	15.3	16.1	9.5
Part 3: Electricity consumption													
Electricity consumption (TWh)	843	926	1,002	1,076	1,128	1,160	1,231	1,347	1,468	1,639	1,892	2,174	2,469
Annual growth rate (%)	11.0	9.9	8.2	7.4	4.8	2.8	6.1	9.5	9.0	11.6	15.5	14.9	13.6
Part 4: Intensity (based on 2000 constant price)													
Energy intensity													
Original energy/original GDP (tce/Y 1,000)	0.240	0.226	0.218	0.211	0.192	0.171	0.157	0.146	0.140	0.142	0.150	0.158	0.162
Original energy/revised GDP (tce/Y 1,000)	0.222	0.207	0.200	0.192	0.175	0.155	0.142	0.131	0.126	0.126	0.133	0.139	0.143

(continued)

	1993	1994	1995	1996	1997	1998	1999	2000	2001	2002	2003	2004	2005
Part 4: Intensity (based on 2000 constant price) (continued)													
Adjusted energy/original GDP (tce/Y 1,000)	0.240	0.226	0.218	0.211	0.192	0.171	0.162	0.155	0.149	0.146	0.154	0.163	0.162
Adjusted energy/revised GDP (tce/Y 1,000)	0.222	0.207	0.200	0.192	0.175	0.155	0.146	0.140	0.133	0.129	0.136	0.143	0.143
Electricity intensity													
Electricity consumption/original GDP (kWh/Y)	0.174	0.170	0.167	0.163	0.157	0.150	0.149	0.151	0.153	0.157	0.166	0.174	0.180
Electricity consumption/revised GDP (kWh/Y)	0.161	0.156	0.153	0.149	0.143	0.136	0.134	0.136	0.137	0.140	0.147	0.153	0.158
Part 5: Elasticity													
Original energy vs. original GDP	0.46	0.46	0.65	0.62	−0.09	−0.52	−0.22	0.02	0.47	1.19	1.65	1.60	1.31
Original energy vs. revised GDP	0.45	0.44	0.63	0.59	−0.09	−0.52	−0.21	0.02	0.43	1.08	1.53	1.51	1.31
Adjusted energy vs. original GDP	0.46	0.46	0.65	0.62	−0.09	−0.52	0.17	0.44	0.45	0.72	1.64	1.70	0.96
Adjusted energy vs. revised GDP	0.45	0.44	0.63	0.59	−0.09	−0.52	0.16	0.42	0.40	0.66	1.53	1.60	0.96
Electricity vs. original GDP	0.82	0.78	0.78	0.77	0.55	0.36	0.86	1.19	1.20	1.40	1.66	1.57	1.37
Electricity vs. revised GDP	0.79	0.75	0.76	0.74	0.52	0.36	0.80	1.13	1.08	1.27	1.55	1.47	1.37

Source: Development Research Center. Various years. *China Energy Statistical Yearbook.* Beijing: China Statistics Press.

Note: — = not available; GDP = gross domestic product; kWh = kilowatt-hour; tce = tons of coal equivalent; TWh = terawatt-hour.

APPENDIX B

Biomass Energy Use in China

Summary

Biomass energy (firewood and straw) is an important fuel for rural residents in China. It is significant at the national level: biomass use approaching 300 million tons of coal equivalent (mtce) amounted to about 13 percent of total primary energy consumption from 2000 to 2004. The development and use of biomass energy, using modern technologies, is an important way to improve the quality of life of rural residents and to reduce environmental pollution. It is also a cost-effective option to reduce the emission of carbon dioxide.

Biomass Resources

China is rich in biomass resources, which are widely distributed across the country. Total biomass resources that can be developed and used as energy amount to about 700 mtce.[i] This figure includes firewood, 120 mtce; straw, 150 mtce; and other sources such as industrial organic wastewater, excrement and urine of humans and animals, urban residential wastewater, and garbage.

Current Status of Development

China is the world's largest biomass energy consumer. In 2002, its biomass energy consumption was 18 percent of the world total (IEA 2004).

Direct Burning

In 2004, total biomass directly burned in the rural areas of China amounted to 299.0 mtce, of which 266.2 mtce was used for cooking, hot water, and heating supply and 32.8 mtce was used by rural industries. To improve the efficiency of biomass energy and reduce the environmental pollution associated with its use, China has been promoting the use of fuel-saving closed stoves for many years. At the end of 2004, 189 million households in rural areas were using these appliances. The thermal efficiency of the fuel-saving stove is more than 25 percent, double the efficiency of conventional stoves.

Biogas

China also takes the global lead in the application of biogas technology and biogas pools. By the end of 2004, 14.46 million biogas pools were established in rural areas, with an annual biogas production of 5.57 billion cubic meters. About 3,090 large and medium-size biogas pools use industrial organic wastewater and excrement from livestock-breeding farms; their annual biogas production is 1.10 billion cubic meters.

Gasification of Straw

By the end of 2004, there were 525 straw-gasification plants, with an annual gas production of 180 million cubic meters.

Generation

In 2004, the installed capacity of biomass-fired generation (mainly bagasse) was 1.9 gigawatts.

Liquid Fuels

In 2004, 800,000 tons of fuel ethanol and 200,000 tons of biodiesel were produced.

Prospects for Biomass in China's Energy Balance

By 2020, according to the China Renewable Energy Development Plan, published by the National Development and Reform Commission in

Table B.1 Development of Biomass Energy in China

Biomass energy	1980	2000	2004	2005	2020 (forecast)
Direct burning (mtce)	229.0	219.1	299.0	295.6	150.0
Share of primary energy consumption (%)	27.5	13.6	12.8	11.7	4.3
Biogas (billion cubic meters)					
Household biogas pool	1.01	2.59	5.57	6.90	15.00
Large and medium-size biogas pools	n.a.	0.52	1.10	1.50	10.00
Generation (gigawatts)	n.a.	0.8	1.9	2.0	30.0
Liquefaction					
Fuel ethanol (million tons)	n.a.	n.a.	0.8	0.8	19.0
Biodiesel (million tons)	n.a.	n.a.	0.2	0.2	1.0

Sources: China Ministry of Agriculture 2005; Editorial Board of the *China Rural Energy Yearbook 1997* 1998; Institute of Nuclear and New Energy Technology and Tsinghua BP Clean Energy Research and Education Center 2005.

Note: n.a. = not applicable. Primary energy consumption includes biomass consumption by direct burning.

June 2005, and according to projections by local experts, biomass consumption in the form of direct burning will be reduced to 150 mtce, biogas production from the household biogas pools will be 15.0 billion cubic meters, biogas production from industrial organic wastewater and livestock-breeding farms will be 10.0 billion cubic meters, total installed capacity of biomass-fired generation will be 30 gigawatts, and 19 million tons of fuel ethanol and 1 million ton of biodiesel will be produced. These data, together with historical information provided for perspective, are set out in table B.1

Adverse Effects on the Natural Environment and on Human Health from Biomass Use

About 750 million people live in rural areas in China. Because these people have limited access to commercial energy forms and have low incomes, much biomass energy is consumed in these areas. The traditional ways of using biomass energy have already caused serious harm to the eco-environment and to human health.

In 2004, about 56 percent of rural energy for residential use was derived from biomass. The indoor pollution resulting from direct burning of biomass energy is in the same category as the harm resulting from smoke (World Bank 1997). In 2004, total consumption of firewood amounted to 256 million tons. This amount was 30 percent higher than the reasonably sustainable harvest and thus resulted in the serious destruction of forest vegetation and consequent loss of soil. Much of the available straw was burned as fuel and therefore could not be used as fertilizer in farmland. That absence was one of the major causes of lowered organic matter content in the land and worsened soil fertility (Wang 2005).

Policies Designed to Encourage the Use of Modern Biomass Technologies

The government of China has already taken several steps to promote the development of biomass energy using modern technologies as follows:

- The central government will subsidize large and medium-size biogas projects by meeting 50 percent of the interest cost of borrowings. Some provincial governments also provide some subsidies or low-interest loans to rural household biogas projects.
- According to the implementation decree of the Renewable Energy Law, a tariff premium of Y 0.25 per kilowatt-hour is allocated to biomass-fired generation and the incremental cost will be shared nationally. Furthermore, the value added tax for biogas generation is reduced from 17 percent to 3 percent.
- A subsidy is provided for fuel ethanol production from biomass: for each ton of fuel ethanol sales, a subsidy of Y 1,883 was provided in 2005; Y 1,628 will be provided in 2006, and Y 1,373 in 2007 and 2008.

Note

1. The amounts of biomass resources referred to in this appendix were calculated by Gu Shuhua, a professor at Tsinghua University.

References

Editorial Board of the *China Rural Energy Yearbook 1997*. 1998. *China Rural Energy Yearbook 1997*. Beijing: China Agricultural Press.

China Ministry of Agriculture, Department of Science and Education. 2005. *China Rural Renewable Energy Statistics 2005*. Beijing: China Ministry of Agriculture.

IEA (International Energy Agency). 2004. *World Energy Outlook, 2004.* Paris: IEA.

Institute of Nuclear and New Energy Technology and Tsinghua BP Clean Energy Research and Education Center. 2005. *Proceeding of the China Renewable Energy Development Strategy Workshop.* Beijing: Tsinghua University. http://www.martinot.info/China_RE_Strategy_Proceedings.pdf.

Wang, Qingyi. 2005. "Energy and Environment in China: Problems and Solutions." *Energy and Environment* 3: 4–11.

World Bank. 1997. *Clear Water, Blue Skies.* Washington, DC: World Bank.

APPENDIX C

The Chinese System for Energy Statistics
History, Current Situation, and Ways to Improve the System

The Groundwork Established in the 1980s

In 1982, a special energy statistics institution was established within the China National Bureau of Statistics (NBS). In 1985, the system of energy statistics was set up. It included policies and procedures for surveys of energy consumption and production; for surveys of the energy intensity of industrial products; for data submissions by major energy-intensive enterprises regarding their energy purchases, consumption, and storage; and for compilation of regional energy balances and their submission to the NBS. Enterprises submitted the statistical tables to the corresponding government agencies and statistical bureaus. The NBS then gathered the tables for later processing and publication. The NBS and related government agencies also carried out the sampling of the data that had been submitted. By the end of the 1980s, therefore, China had already established a preliminary energy statistical system that produced comprehensive national and regional energy balances, energy supply and demand projections, and data on the cost-effectiveness of energy use.

Retrenchment in the 1990s

Unfortunately, the government institutional reforms of the 1990s seriously weakened the system for energy statistics. At present, only four staff members in NBS work on energy statistics, focusing mainly on energy consumption data. At the provincial level, only half of the provinces have even one full-time staff member working on energy statistics. As a result, the energy statistics function is weak, the necessary surveys have been cut back, the quality of the published statistics has deteriorated, and—especially troubling—some false statistical data have appeared. This situation often attracts criticism—both in China and internationally.

The Current Situation and Its Weaknesses

Apart from the surveys of energy production and consumption made every five years, other surveys currently include energy production; energy consumption of industrial enterprises with annual sales income higher than Y 5 million; energy consumption, processing, and conversion for enterprises whose annual energy consumption is greater than 10,000 tons of coal equivalent (tce); and a survey of residential energy consumption in cities. The energy consumption of small enterprises (those with annual sales income of Y 5 million or less) is normally estimated. No special investigations are made into such matters as energy consumption in the agriculture, construction, road transportation, and commercial sectors. Information is also lacking on energy supply and demand in spot markets, energy tariffs, energy and the environment, and so on. The survey of the energy intensity of industrial products was discontinued in 1994, and the requirement to compile and submit the regional energy balances to the NBS was eliminated in 1998. China's current energy statistics activity therefore focuses only on the comprehensive energy balance. The statistical functions required to reflect such important features of the economy as energy market supply and demand and the cost-effectiveness of energy use and its environmental impact are missing. Moreover, because much estimation under the current energy statistics system is not supported by proper surveys, there is, unfortunately, a degree of arbitrariness in the system, reflecting the introduction of human judgment (Beijing Energy Efficiency Center 2002)

The design of the energy indices is relatively dated and is based on concepts that are very different from common international definitions (Zhu 2005). This problem creates difficulties for the system of energy statistics itself, as well as for international comparisons and communications.

For example, the final energy consumption in Chinese energy balances includes consumption by the energy sector, which is not the international norm. Nevertheless, in tables C.1 and C.2 (see p. 196) some statistics are presented that attempt to compare Chinese statistics with data based on international definitions. Table C.3 (see p. 196) shows the conversion factors.

In Chinese statistics, sector categories are sorted by their institutional arrangements (for example, ownership of the enterprise) instead of being classified by the sector's activities (for example, road transportation). So petroleum consumption reported for the road transportation sector covers vehicles owned by the public sector only, not those owned by other sectors and individuals. However, in 2003, the gasoline consumption by vehicles owned by the public sector only amounted to about 35 percent of total gasoline consumption of all vehicles. Another example concerns the construction materials sector. According to information provided by the China Building Materials Industry Association, the energy consumption by enterprises that are state owned or managed by central or provincial governments amounted only to about 56 percent of the energy consumption by all construction materials enterprises (both state owned and private). Furthermore, the energy data for each sector are not classified by type of fuel-using equipment or by fuel type, which makes analyzing energy efficiency and related environmental pollution and carbon dioxide emissions difficult.

Conclusions and Recommendations

Serious problems are affecting China's energy statistical system. The energy statistics information system is weak. The design of the system lacks a proper rationale, and the energy indices are therefore deficient. The growing number of enterprises that are not state owned means that sector data collected from only publicly owned or managed enterprises will increasingly be incomplete. A reluctance to share energy statistics is present. The credibility of China's energy statistics is undermined by unsound systems for information collection, analysis, and processing; by insufficient communication among government departments; by attempts to protect local benefits; and by the existence of faked, conflicting, and confusing energy statistics data resulting from arbitrary human judgments. The outstanding recent example of failure in the statistics system is coal production and consumption: the revised numbers for raw coal consumption in 2000 are 30 percent higher than the originally published data.

Table C.1 Final Energy Consumption by Sector, 1980–2000
mtce (percentage of total final energy consumption)

Sector	1980			2000			
	Chinese statistics		Correction based on international definitions	Chinese statistics		Correction based on international definitions	International statistics
	Coal equivalent calculation method	Calorific value calculation method		Coal equivalent calculation method	Calorific value calculation method		
Agriculture	34.7 (6.0)	24.7 (5.2)	31.4 (7.0)	60.5 (4.6)	42.9 (4.4)	40.2 (4.9)	—[a]
Industry[b]	392.1 (68.3)	308.1 (65.2)	277.0 (61.3)	914.1 (69.2)	664.7 (68.5)	489.2 (59.2)	494.3 (60.0)
Transportation[c]	29.0 (5.0)	27.6 (5.8)	36.0 (8.0)	97.2 (7.4)	89.8 (9.2)	134.8 (16.3)	121.4 (14.7)
Residential, commercial, and other sectors[d]	119.3 (20.7)	112.3 (23.8)	107.3 (23.7)	248.5 (18.8)	173.7 (17.9)	162.5 (19.6)	208.6 (25.3)
Total	575.1 (100.0)	472.7 (100.0)	451.7 (100.0)	1,320.3 (100.0)	971.1 (100.0)	826.7 (100.0)	824.3 (100.0)

Sources: The 2000 Chinese statistics are revised energy data provided by the China National Bureau of Statistics. The 1980 Chinese statistics using the coal equivalent calculation method were provided by the China National Bureau of Statistics. The 1980 data using the calorific value calculation method were calculated by the authors. International data were provided by the International Energy Agency.

Note: As noted in the main text, the focus of this report is on commercial energy sources. Biomass energy is not included in these data. Biomass energy forms contribute about 13 percent of China's consumption of all sources of energy. Further information on biomass energy forms, prospects, and issues is found in appendix B.

a. The agriculture sector is included in "other sectors" in the IEA statistics.

b. The data for the construction sector are included in the industry sector.

c. The data for the transportation sector also include energy consumption by the storage, post, and telecommunication sectors.

d. Other sectors include other service sectors except for transportation and the national defense sector.

Table C.2 Final Energy Consumption by Fuel Type, 1980–2000
mtce (percentage of total final energy consumption)

Fuel type	1980 Chinese statistics[a]	1980 Correction based on international definition	2000 Chinese statistics[b]	2000 Correction based on international definition	International statistics
Solid fuel[c]	335.7 (58.4)	311.7 (69.0)	421.6 (43.4)	382.1 (46.2)	355.7 (43.2)
Liquid fuel	89.6 (15.6)	82.1 (18.2)	267.9 (27.6)	213.2 (25.8)	274.3 (33.3)
Gaseous fuel[d]	23.3 (4.0)	19.1 (4.2)	78.3 (8.1)	66.7 (8.1)	25.7 (3.1)
Electricity	116.2 (20.2)	28.3 (6.3)	154.1 (15.9)	128.3 (15.5)	132.9 (16.1)
Heat	10.3 (1.8)	10.5 (2.3)	49.2 (5.0)	36.4 (4.4)	35.7 (4.3)
Total	575.1 (100.0)	451.7 (100.0)	971.1 (100.0)	826.7 (100.0)	824.3 (100.0)

Source: International data are provided by the International Energy Agency. Chinese data are provided by the China National Bureau of Statistics.

a. Statistics are based on the coal equivalent calculation method. The 1980 final energy consumption data is not available based on the calorific value calculation method.

b. Statistics are based on the calorific value calculation method.

c. Solid fuel includes coal, briquettes, and coke.

d. Gaseous fuel includes coal gas, liquefied petroleum gas, and natural gas. However, in International Energy Agency statistics, liquefied petroleum gas is classified as a liquid fuel.

Table C.3 Conversion Factors for Tables C.1 and C.2

Fuel	Conversion factor
Raw coal	0.7143 kilogram coal equivalent per kilogram
Coke	0.9714 kilogram coal equivalent per kilogram
Crude oil	1.4286 kilogram coal equivalent per kilogram
Gasoline and kerosene	1.4714 kilogram coal equivalent per kilogram
Diesel oil	1.4571 kilogram coal equivalent per kilogram
Fuel oil	1.4286 kilogram coal equivalent per kilogram
Liquefied petroleum gas	1.7143 kilogram coal equivalent per kilogram
Natural gas	1.33 kilogram coal equivalent per cubic meter
Heat	0.03412 kilogram coal equivalent per megajoule
Electricity:	
Calorific value method	0.1229 kilogram coal equivalent per kilowatt-hour
Coal equivalent method (1980)	413 gram coal equivalent per kilowatt-hour
Coal equivalent method (1990)	392 gram coal equivalent per kilowatt-hour
Coal equivalent method (2000)	363 gram coal equivalent per kilowatt-hour
Coal equivalent method (2004)	349 gram coal equivalent per kilowatt-hour

Source: China National Bureau of Statistics and National Development and Reform Commission 2005.

However, the Chinese energy statistics system is being improved. The government of China is taking some significant measures to reform its statistics management institution and to improve the statistical system. The State Council issued the revised version of the Implementation Decree for the Statistics Law on December 16, 2005, and the decree came into effect on February 1, 2006. To better ensure the reliability of statistical data, the NBS now leads the teams responsible for the rural social economy survey, the urban social economy survey, and the enterprise survey. The NBS provides vertical coordination to carry out the surveys and prepare reports and generally supervises the work. The survey teams are also authorized to check that the Statistics Law is properly implemented and to penalize those who disobey the law.

The following steps would improve the energy statistics system:

- Enhance capacity building in energy statistics.
- Establish the energy statistics indices system on a basis consistent with international definitions and practices.
- Implement the energy statistics reporting system for all major energy-intensive enterprises.
- Improve the energy statistics survey system with the NBS carrying out the survey of enterprises directly, using modern information technology systems.
- Improve the energy balance statistics to include the cost-effectiveness of energy use and to present the energy-environment relationship.
- Establish a transparent, timely, nondiscriminatory energy statistics information publication system.
- Establish the national energy statistics multipurpose electronic information system and database.

References

Beijing Energy Efficiency Center, Industry and Transportation Department, China National Bureau of Statistics. 2002. *Study on the Improvement of Chinese Energy Statistics, Indices, and Methodology.* Beijing: China National Bureau of Statistics.

China National Bureau of Statistics and National Development and Reform Commission. 2005. *China Energy Statistical Yearbook 2004.* Beijing: China Statistics Press.

Zhu, Hong. 2005. "Improve the Statistic of Energy Balance to Adapt to the Scientific Development Requirement." *Energy of China* 7 (27): 16–20.

APPENDIX D

Energy Costs as a Proportion of Gross Domestic Product

Estimates for China, Japan, and the United States

The main text of this report presents estimates for 2005 of energy costs as a proportion of gross domestic product (GDP) as follows: China, 13 percent; Japan, more than 3 percent; and the United States, about 7 percent.

The assumptions for energy consumption, for unit energy costs and values, and for GDP underlying these estimates are set out in tables D.1, D.2, and D.3 (see p. 200) and in the notes to those tables.

These estimates of what might be termed *raw energy input costs as a proportion of GDP* are put forward to make international comparisons based on orders of magnitude only. It is accepted that there may be grounds to challenge the details of the unit energy costs used to derive the estimates.

To the extent that the oil prices used may be high for 2005 (they are low in relation to costs experienced in the first half of 2006), their use will tend to exaggerate the estimated gross energy costs of Japan and the United States relative to China.

Note that these estimates relate to the costs of the gross energy inputs to the respective economies valued at approximate market prices. This approach is used even though some consumers may not be paying market prices (for example, some gas consumers in China pay "in-plan" prices that may be well below market value).

Table D.1 Total Primary Energy Consumption and Fuel Shares in China, 2005

Fuel type	Amount consumed (mtce)	Price (US$ per tce)	Total fuel share (US$ billion)
Coal	1,529	50[a]	76
Oil	472	302[b]	142
Gas	62	196[c]	12
Hydropower and nuclear	162	50[d]	8
Total	2,225	n.a.	238 (12.9% of GDP)[e]

Source: Compiled by the authors using basic consumption data from table 2.6.

Note: mtce = million tons of coal equivalent; tce = tons of coal; n.a. = not applicable.

a. Price is based on the Quinhuangdao trading center price at mid 2005. It may be on the high side, but taking off about US$10 per tce will not make a significant difference to the final energy-to-GDP ratio.

b. Price is based on the Dubai crude oil spot price of US$50 per barrel plus US$2 for transportation, US$2 for the refining margin, and US$5 for distribution and so forth.

c. Gas is valued at 65 percent of oil.

d. Primary electricity is expressed in fuel equivalence, so a coal price of US$50 per tce was assumed.

e. We assume a GDP of US$1.848 trillion, which equals the most recent GDP data for 2004 published by the China National Bureau of Statistics increased by 8 percent.

Table D.2 Total Primary Energy Consumption and Fuel Shares in Japan, 2005

Fuel type	Amount consumed (mtce)	Price (US$ per tce)	Total fuel share (US$ billion)
Coal	182	50	9
Oil	366	302	111
Gas	114	196	22
Hydropower and nuclear	129	50	6
Total	791	n.a.	148 (3.4% of GDP)[a]

Source: BP Global 2006.

Note: mtce = million tons of coal equivalent; tce = tons of coal equivalent; n.a. = not applicable. Data are converted from tons of oil equivalent to tons of coal equivalent at a ratio of 1.5. The assumed per unit energy prices are the same as for China in table D.1

a. Japan's GDP is US$4.301 trillion.

Table D.3 Total Primary Energy Consumption and Fuel Shares in the United States, 2005

Fuel type	Amount consumed (mtce)	Price (US$ per tce)	Total fuel share (US$ billion)
Coal	855	40[a]	34
Oil	1,417	336[b]	476
Gas	861	237[c]	204
Hydropower and nuclear	370	50[d]	19
Total	3,503	n.a.	733 (6.7% of GDP)[e]

Source: BP Global 2006.

Note: mtce = million tons of coal equivalent; tce = tons of coal equivalent; n.a. = not applicable. Data are converted from tons of oil equivalent to tons of coal equivalent at a ratio of 1.5.

a. This price is derived from a mix of spot prices for 2005 focusing on the Illinois Basin and Central Appalachia: coal in the United States is very cheap. Even so, these numbers may be on the high side.

b. This price is based on the West Texas Intermediate crude oil average of US$56.59 per barrel, with allowance for transportation, refining margin, marketing, and distribution. U.S. oil supply costs are probably marginally higher than China's.

c. This price is based on the average city gate price for 2005 according to the U.S. Department of Energy, which happens to be 70 percent of the assumed oil price, compared with 65 percent for China.

d. BP Global (2006) uses the fuel equivalent method of expressing primary electricity's contribution.

e. The U.S. GDP is US$10.949 trillion.

Moreover, the estimates do not measure the costs to energy consumers of commercial energy at the meter (in the cases of gas and electricity) or at the retail pump (in the case of automotive fuels) or of coal delivered to a consumer's industrial plant or residence. The estimates do not cover the costs of conversion to electricity, nor do they include taxes on energy production, conversion, or use that may fall on final consumers and be reflected in the prices they pay.

Reference

BP Global. 2006. *Statistical Review of World Energy 2006*. London: BP Global. http://www.bp.com/productlanding.do?categoryId=91&contentId=7017990.

APPENDIX E

Feedback from the Dissemination Workshop

On June 1–2, 2006, a workshop was held in Beijing to disseminate and discuss this report, which was then in late-draft form. The purpose was to obtain feedback from a broad spectrum of Chinese government officials, energy industry representatives, and academics. A number of senior foreign experts also took part. The authors took careful account of the valuable views expressed and incorporated them appropriately in the final text.

Feedback 1: China's Energy Economy Presents Many Challenges

While acknowledging the achievements of the past, Chinese participants at the workshop almost unanimously backed the report's evaluation of the sector and emphasized that the nation's energy economy faces many challenges. Some examples mentioned include the following:

- The energy statistical system is not reliable and requires a comprehensive overhaul.
- Energy consumption per unit of output is still high.
- Gas industry growth has been slow.
- Energy technology is relatively backward in some subsectors.

- The government's management of the sector has not been efficient. In particular, institutional development still lags the nation's needs.
- The roles of government and industry have not been properly defined.
- Strong government guidance of the sector is needed and, implicitly, has not yet been provided.

Some paradoxical remarks during the workshop highlighted the hybrid nature and middle-of-the-stream status of the sector. Most of the foreign participants emphasized government intervention, and a number of Chinese participants stressed the need for increased reliance on the market. Such remarks are a troubling indication that the hybrid sector is not getting the best elements of the two systems. A consensus does exist that the present path will not achieve sustainability.

The report stresses that weaknesses also present opportunities:

- Low per capita energy consumption and deficiencies in energy efficiency increased the awareness of the need to use advanced technologies to meet the fast-growing demand. In this way, China may be able to avoid the highly energy-intensive development paths followed by even the most energy-efficient economies.
- Rapidly increasing oil imports put energy security and safety on the government's high-priority list even though the country's energy dependence is quite low.
- Excessive reliance on coal and ensuing pollution are great incentives for China to deploy the most advanced clean coal technologies and to become a leader in the field.

Feedback 2: The Energy Statistical System Needs Revision

Chinese interlocutors agreed that there are weaknesses in the energy statistical system that still must be filtered out. They expressed concern about the accuracy of data and that some parties may be inclined to play the numbers game, especially in relation to the 20 percent energy-intensity reduction during the 11th Five-Year Plan (2006–10). So far, the results of work by the provinces to break out consumption data by sector have been uneven. Data are not available on the extent to which growth in the consumption of commercial energy reflects a switch from the use of traditional biomass in rural areas.

The report stresses that attention must be given to improving historical energy data and establishing comprehensive data flows. The current data situation is unsatisfactory; data are often unreliable and are subject to revisions long after the fact. The report recommends developing a strong energy statistical system that can produce reliable energy balances and other information on a timely basis. Appendix C helps address the issues discussed during the workshop.

Linking the performance of local government departments to the achievement of consumption targets of the 9th Five-Year Plan (1996–2000) has likely contributed to the minimization and distortion of the energy consumption statistics during the late years of the 9th Five-Year Plan. These distortions undermined the credibility of the energy statistical system. China still has some way to go in achieving a reliable energy statistical system.

Elsewhere, progress is being made. In the United States, for example, critical energy supply data are published on a weekly basis. In Canada, energy consumption statistics are available for seven categories, for five fuels, and for seven regions. Each of the consumption categories is further subdivided into as many as 10 groups, therefore yielding hundreds of energy data points.

Feedback 3: The Target 20 Percent Energy-Intensity Reduction during the 11th Five-Year Plan Will Be Difficult to Achieve

The workshop participants saw achievement of the target 20 percent energy-intensity reduction as a big challenge and expressed a less optimistic view than the report for achieving it. Officials are still identifying the expected sources of efficiency gains, statistical baselines have not yet been completed, measures and regulations to achieve the target have not been promulgated, local governments may not yet have signed on to the 20 percent goal, and sectoral and regional targets have not yet been assigned. Meanwhile, energy over gross domestic product (GDP) elasticity continues to be high, and the goal of a 4 percent reduction in the first year of the 11th Five-Year Plan is unlikely to be achieved.

Participants seemed discouraged by the enormity of the task and the apparently slow progress being made in addressing it. Such discouragement must not be allowed to detract from the great value and central importance of the target in relation to the key sustainability objective of reining in energy consumption. The report states that the 20 percent intensity reduction is a feasible but daunting challenge that requires an aggressive, comprehensive implementation program and supporting policies and incentives. The 20 percent figure is endorsed as a stretched target and is meant to provide a rallying point for the effort to achieve long-term sustainability.

Feedback 4: Economic Well-Being, Energy Consumption, and Environmental Concerns Should Be Addressed

Participants recognized, and the report acknowledges, that China is still a relatively poor country that understandably aspires to the living standards of the advanced industrial states. Against that background, some of the participants questioned, "Is the report saying that the Chinese people cannot reach those living standards because they cannot be 'allowed' to consume energy on a scale relative to that consumed by the populations of those other countries?"

Responsive to this concern, the report is clear that energy policy should (and can) achieve multiple objectives simultaneously. It must not compromise either the continued strong economic growth needed to support rising living standards for China's people or the environmental progress needed to improve quality of life and save the country's limited resources of air, water, agricultural land, forests, and uncongested space.

Although the report warns about the risks for the energy economy and the environment of North American–style (and, to a lesser extent, European-style) suburban living and mass motorization, its clear message is that energy policies *can* be shaped to support high growth rates and rising living standards with much less energy use per unit of output and improved environmental results.

In the context of environmental concerns and progress, participants accepted that climate change is a particularly serious problem for China and that the country therefore has a strong vested interest in doing something about the problem.

On a per capita basis, China's energy consumption and atmospheric emissions are still low by international standards. But in the aggregate, they are already large and, more important, their volume growth is huge, sustained, and unprecedented. They are seriously harming China's environment and are becoming a major concern on a global scale. They are forecast to account for 25 percent of greenhouse gas emissions worldwide over the next two decades. For both reasons, vigorous efforts are needed to rein in consumption and properly integrate energy and environmental policy. Failure to do so will constrain economic growth and the Chinese people's well-being.

Feedback 5: Energy Technology Leapfrogging Requires Clarification

Participants debated the definition of *leapfrogging*, a concept that has been used in a wider social development context. Some participants indicated that leapfrogging could be applied to economic development models (phases) but were uncertain about which international experience is most relevant for China. It was clarified that the concept is used in a more restrictive sense in the report even if the tunnel effect (that is, avoiding the high-energy and GDP elasticity phase) is a leapfrogging in the broad economic development sense. Two definitions encapsulate the use of the concept in the report:

- *Leapfrogging* is "a theory in which developing countries skip inferior, less efficient more expensive, or more polluting technologies and industries and move directly to more advanced ones" (Wikipedia Foundation 2006).
- *Leapfrogging technologies* refers to "the use of technology in developing economies to find new paths to development, skipping the slower route that other nations took. Innovation is the cornerstone of leapfrogging technologies. It is not just about identifying technologies but also about finding new ways to apply existing ideas in a different context where, for example, the existing infrastructure may be very limited [or, in the context of this report, growing very fast]. Leapfrog technologies are, more often than not, those which do not require an existing grid—mobile phones, wireless communications, distributed power supply like power solar, and so on." (PR Leap 2005).

This report extends that concept to fast-growing subsectors such as power generation, where the use of clean coal technologies could achieve the outcome of rapid expansion of electricity supplies more efficiently and with less pollution than could conventional coal-thermal technology.

Some participants stated that there is huge potential for energy technology advancement and great opportunities for leapfrogging. Also, some existing technologies are not being properly used. Technical leapfrogging can be encouraged by establishing and enforcing efficiency standards, a necessary form of command and control; by focusing on advanced technologies such as third-generation nuclear plants; and by harnessing the resources of both state-owned enterprises and the venture capital sector. Participants also talked about the extent to which economic structure change can reduce energy intensity of the economy. At the same time, they recognized that structural change will take time and that vested interests may be an obstacle. Examples were given of energy-intensive and polluting factories that have been attracted to some localities by tax and labor market incentives.

Other participants expressed skepticism about leapfrogging and felt the need for clear ideas about what can and cannot be leapfrogged. Some indicated that government must play a dominant role in undertaking projects. Others questioned why China, still a poor country, should pay for the deployment of higher-cost technologies that are not even deployed in more advanced and richer countries.

The report stresses that leapfrogging to the technological leading edge and avoiding energy-intensive development models (for example, urbanization requiring individual car ownership, rather than efficient public transportation, to meet mobility needs and inefficient architectural design posing excessive heating and cooling energy needs) are necessary conditions for China to achieve energy sustainability. Without them, further large efficiency gains will be difficult to achieve. China is better positioned to deploy these technologies than many industrial countries because of its market size and because it has fewer vested interests and stranded costs, which tend to delay the development of these technologies in more advanced countries. Backwardness is a great opportunity for technological leapfrogging. Proper incentives, technology transfer, international cooperation, and participation in the Clean Development Mechanism (CDM) market under the Kyoto Protocol are identified as some of the conditions for success. These conditions could combine to make China's development model radically different from the models followed by industrial countries. All international experiences are relevant to China to avoid energy-intensive paths and devise a different development model. The government of China can probably best play a mainly supportive role by getting market signals right, ensuring that externalities are reflected in investment decisions, providing capital assistance where economics are marginal, and perhaps underwriting demonstration projects.

Feedback 6: International Cooperation and Technology Transfer Are Needed

Chinese participants agreed that international energy cooperation in many fields is key in such areas as technology transfer and security of energy supply and in relation to global climate change issues. The willingness of industrial countries to provide access to the most advanced technologies was also questioned. There were calls for the institutionalization of this function within government.

Broadened international cooperation with mutual benefits to all stakeholders is one of the report's building blocks for a sustainable energy policy, with particular regard to the security and transfer of leading-edge technologies, which is presented as another building block. Related to the latter building block, an important role is foreseen for the use of revenue streams, available under the Kyoto Protocol's CDM, to fund such transfers. It is implicitly recognized that countries such as China and India, where the potential of deploying clean coal technologies exists, cannot afford to access these technologies on a commercial basis only. Is the world doing enough? That is the central question that needs to be answered quickly so as to develop adequate strategies to address the increasing negative impacts of climate change.

Given the size and dispersion of China's current international energy cooperation activities, their prospective growth, and their importance for sustainability, consideration could be given to creating an institutional framework, at least to the extent of a clearinghouse for gathering, collating, and disseminating information about international activities and their results. As for whether China should play an active or a passive role in relation to CDM, the market for emissions reductions that China offers is surely large enough that it could play an active role in shaping the use of CDM within its established principles.

References

PR Leap. 2005. "Leapfrogging Technologies: A Different Route to Development." News release, September 29. http://www.prleap.com/pr/15733.

Wikipedia Foundation. 2006. "Leapfrogging." http://en.wikipedia.org/wiki/Leapfrogging.

APPENDIX F

Life-Cycle Costs of Electricity Generation Alternatives with Environmental Costs Factored In

The levelized cost of an integrated coal gasification combined cycle (IGCC) unit was analyzed to provide a basis for comparison with a conventional subcritical 600-megawatt coal-fired generator with flue gas desulfurization (FGD), which is the predominant kind of new thermal generation capacity being added in China. Table F.1 summarizes the main technical characteristics of an IGCC unit and a subcritical 600-megawatt coal-fired thermal generator with FGD.

Three cases were analyzed: (a) without environmental externality, (b) with environmental externality and the carbon credit valued at US$6 per ton of carbon dioxide, and (c) with environmental externality and the carbon credit valued at US$12 per ton of carbon dioxide. The difference in local emissions—total suspended particulates (TSP), sulfur dioxide, and nitrogen oxides—between the IGCC unit and the 600-megawatt coal-fired thermal generator was also considered, although this difference is much smaller than the carbon emissions difference. The externality costs of TSP, sulfur dioxide, and nitrogen oxides were valued at US$2,488 per ton, US$525 per ton, and US$625 per ton, respectively, on the basis of a benefit transfer method for eastern China.

Table F.1 Main Technical Indices of IGCC and Subcritical 600-Megawatt with FGD Units

Indices	IGCC	Subcritical 600-MW coal with FGD
Investment (US$/kW)	1,038.00[a]	562.40
Fixed cost (US$/kW)	22.50	16.80
Variable cost (mills/kWh)	0.1	0.714
Auxiliary consumption rate (%)	14.5	5.6
Gross unit efficiency (%)	52.8	41.0
Coal consumption rate (gce/kWh)	233	300
Fuel price[b] (Y/tce)	650	650
Annual use (hours)	7,000	7,000
Economic life (years)	25	25
Discount rate (%)	12	12
Construction period (years)	4	5

Source: World Bank 2001.

Note: kW = kilowatt; kWh = kilowatt-hour; MW = megawatt; gce = grams of coal equivalent; tce = tons of coal equivalent.

a. In this analysis, the investment of an IGCC unit is selected as a sensitive variable to check its economic rationality compared with a subcritical 600-megawatt coal-fired thermal unit with FGD.

b. Fuel price is estimated on the basis of the Qinhuangdao free-on-board cost plus transportation fee to eastern China.

The results in panel a of figure F.1 (see p. 210) show that the levelized cost of the IGCC unit can be the same as that of the subcritical 600-megawatt coal-fired thermal plant with FGD in these circumstances:

- The capital cost of the IGCC unit is about 126 percent of the capital cost of the 600-megawatt coal-fired thermal generator—or US$700 per kilowatt or 32 percent lower than its current level—when no environmental externality is considered.
- The capital cost of the IGCC unit can be about 142 percent of the capital cost of the 600-megawatt coal-fired thermal generator—or US$800 per kilowatt or 23 percent lower than its current level—when environmental externality is considered and the carbon credit is valued at US$6 per ton of carbon dioxide.
- The capital cost of the IGCC unit can be about 157 percent of the capital cost of a 600-megawatt coal-fired thermal generator—or US$880 per kilowatt or 15 percent lower than its current level—when environmental externality is considered and the carbon credit is valued at US$12 per ton of carbon dioxide.

Figure F.1 Levelized Cost Comparison of IGCC and Subcritical 600-Megawatt with FGD Units: Capital Cost Ratio

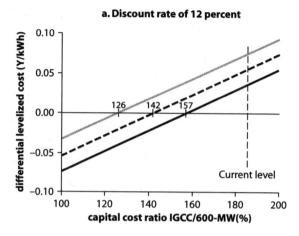

a. Discount rate of 12 percent

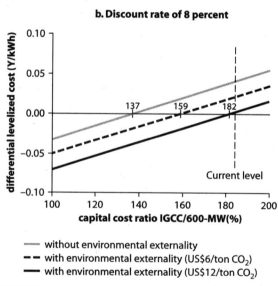

b. Discount rate of 8 percent

— without environmental externality
– – with environmental externality (US$6/ton CO_2)
— with environmental externality (US$12/ton CO_2)

Source: World bank study team.

Note: CO_2 = carbon dioxide, kWh = kilowatt-hour; MW = megawatt.

A sensitivity analysis was conducted with a discount rate of 8 percent, which benefits the more capital-intensive IGCC technology, as shown in panel b of figure F.1.

Figure F.2 presents the same data, except that the x axes shows the change in the assumed capital cost of the IGCC unit rather than, as in

Figure F.2 Levelized Cost Comparison of IGCC and Subcritical 600-Megawatt with FGD Units: Change of Capital Cost of IGCC

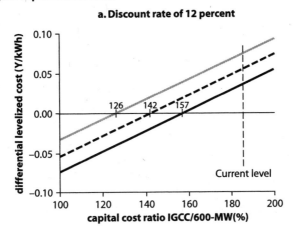

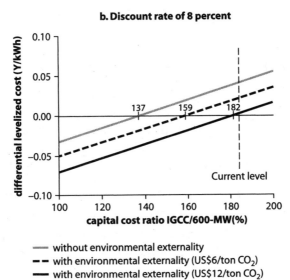

Source: World bank study team.
Note: CO_2 = carbon dioxide, kWh = kilowatt-hour; MW = megawatt.

figure F.1, the ratio of the capital cost of the IGCC unit to the subcritical 600-megawatt unit.

Reference

World Bank. 2001. *Technology Assessment of Clean Coal Technologies for China.* Washington, DC: World Bank.

APPENDIX G

International Experience of Insecurity of Energy Supply

As the main text notes, for the past half-century, international security concerns about energy supply have tended to focus, first, on oil imports. There has been a second, at least latent concern about the security of imported gas supplies. A third concern relates to the importance of electricity security. Moreover, particularly in recent years, there has been a realization that domestic energy sources may not provide truly secure supplies. Coal imports have not attracted significant attention from a security standpoint.

This appendix reviews international experience in relation to each of these areas of potential energy-supply security concerns and, when appropriate, includes assessments and implications for China's energy policies.

Concerns about Oil Imports

Concerns about oil imports reflect the international experience since the middle of the 20th century, during which as many as seven significant oil supply disruptions occurred. There are basically three perceived risks surrounding petroleum imports:

- Major oil-exporting countries, or the associations that group those countries, will change their policies.

- Market failures could result from insufficient investments in exploration, field development, and transport infrastructure.
- *Force majeure* events could result in potential short-term interruptions.

Table G.1 summarizes the main oil supply crises that have taken place since the early 1950s.

Table G.1 Major Global and Regional Oil Supply Crises since 1950

Date	Event
1951–54	The first post–World War II oil crisis occurs: nationalization of the Anglo-Iranian Oil Company by Iranian Prime Minister Mohammed Mossadegh. Exports were cut off for about three years, but additional supplies were quickly developed from other Middle Eastern countries, particularly Kuwait.
1956–57	The Suez Crisis involves the blockage of the Suez Canal, the destruction of pipeline pumping stations, and a resulting global tanker shortage. The United States calls on its spare production to meet the deficit.
1967	The Six-Day Arab-Israeli War causes blockage of the Suez Canal and a global tanker shortage. However, the United States again has sufficient spare capacity to avert a crisis
1973	The Yom Kippur Arab-Israeli War and the Arab oil embargo cause crude oil prices to rise drastically from US$2.90 per barrel in September to US$11.65 per barrel in December. The United States draws down its spare capacity before the war begins. Physical shortages occur in Europe and North America, partly as a result of clumsy interventions by consuming country governments (for example, the United States) and of foreign exchange shortages (for example, in Turkey)
1979–81	During this period, several events cause a dramatic increase in oil prices: Iranian worker strikes, revolution, the hostage crisis, and war with Iraq. Panic sends official oil prices from US$13 per barrel to US$34 per barrel.
1990–91	Iraq invades Kuwait, and Coalition forces invade Iraq, resulting in the loss of exports from both countries, a brief spike in oil prices, and then a fallback because there is enough spare capacity in the world supply system to cope with the export loss.
1990–2003	The United Nations imposes a partial embargo on Iraqi oil exports.
December 2002–January 2003	Strikes and demonstrations in the República Bolivariana de Venezuela cause exports to cease.
2003	The invasion of Iraq, led by the United States, causes a brief curtailment of oil exports, followed by their restoration to near prewar levels. Iraqi production of about 2.0 million barrels a day (exports 1.5 million) in mid 2006 accounts for only about 7 percent of production by the Organization of Petroleum Exporting Countries.

Source: Authors' compilation.

Potential Policy Changes of Major Countries or Country Groupings

Of the individual oil-producing countries, Saudi Arabia has by far the largest oil reserves, the biggest developed capacity to produce crude oil, and the greatest amount of spare producing capacity to deal with supply shortfalls. Its development and export policies are therefore potentially critical for global supply. Of the intergovernmental energy organizations, only the Organization of Petroleum Exporting Countries (OPEC) is in a strong position to have an adverse impact on oil supply. Its members account for about three-fourths of the world's reserves, more than 40 percent of the world's current oil supply, and virtually all of the spare capacity.

Neither Saudi Arabia nor OPEC has signaled any fundamental policy change. Both appear committed to supplying sufficient oil to balance the world's needs. However, structural changes in the industry seem to indicate that this commitment will come at much higher price levels than seemed possible five years ago.

The Organization of Arab Petroleum Exporting Countries (OAPEC), another intergovernmental organization, applied selective export embargoes on crude oil exports to Germany, the United Kingdom, and the United States in 1973 and 1974. However, the organization has renounced the use of the "oil weapon" in support of its members' political aims. OAPEC now devotes its efforts exclusively to commercial and technical cooperation in petroleum matters.

The current situation shows no major policy change, but the concentration of petroleum resources in countries that are members of OPEC gives OPEC the ability to have a substantial impact on the security of global oil supply. OPEC members could limit petroleum production, and reorient bilateral trade relationships.[1] Reductions of output could cause worldwide price increases (mainly for oil) as prices rise to equilibrate curtailed supply against demand. The duration of the effects could be from short term (six months to a year) to long term (more than five years). The importance is obvious for China to maintain good and close diplomatic and trade relations with individual major exporting countries and with their international organizations.

Market Failure Resulting from Underinvestment in Securing Supply

Oil is such an important source of revenue in most oil-exporting countries that it usually comes under the control of governments through national oil companies (NOCs). NOCs do not operate on commercial and profit motives alone. Often they have difficulty accessing investment funds needed to develop the petroleum resources available to them.[2] By contrast, international oil companies (IOCs) are typically well financed. They are more likely to be constrained by investment opportunities rather than by lack of funds, at least in the upstream (production) subsector of the industry.

Industry failure to make needed investments may also result from policies of governments around the world that, for environmental or social reasons, limit the industry's access to geologically attractive areas for petroleum exploration. For example, the U.S. government does not allow new exploration off the east and west coasts of the country. There is also an ongoing dispute about providing oil companies access to strongly prospective areas in the Alaska National Wildlife Reserve. Regulatory delays, sometimes reflecting environmental opposition to project development, can affect the construction of petroleum facilities such as liquefied natural gas (LNG) import terminals, pipelines, and oil refineries in industrial countries. In North America, Western Europe, and Japan, it is very difficult and expensive to find sites for large new plants of this sort. Regulatory approvals for site development are also costly and time consuming. For example, in the United States it takes at least two and one-half years to obtain all the permits needed for a new LNG import terminal.

If market failures limit available supply, the flow of oil may be unable to meet demand. In a free market, prices will rise to keep the market in equilibrium. If prices are not allowed this function, however, physical shortages may result. The duration of these effects could be short term to several years. Box G.1 (see p. 216) reviews the significance of the current oil price spike in the security-of-supply debate.

Expert opinions differ on the future supply and price of oil. Some say that reserves have been increasing, that there is a satisfactory (high) ratio of reserves to production, and that foreseeable supply is sufficient to meet demand to 2020 and beyond. However, others claim that some producing countries and IOCs are overstating their reserves. They also state that member countries of OPEC have not made sufficient investment to turn OPEC reserves into productive capacity; that there are few large, new discoveries being made; that IOCs are having great difficulty replacing their own current production; and that IOCs have achieved only minor production increases in the past decade.

It is widely believed that the global oil industry *may* have entered a new phase in the first years of the 21st century. This phase has two principal features: (a) the demand shock of increasing imports by China and India and the still-growing consumption in the United States and (b) the

> **Box G.1**
>
> ### The 2003–06 Oil Price Spike—How Significant Is It in the Security Debate?
>
> **Price Behavior**
>
> International crude oil prices have been rising irregularly for the past three or four years. At times, they have approached US$80 per barrel. These are the highest nominal levels ever seen. However, they are still less in constant dollar terms than they were in the early 1980s, when, for a time, spot prices of Arab light crude oil reached US$40 per barrel, equivalent to more than US$85 per barrel in 2006. Despite current high oil prices, global economic growth has continued, notably in both China and the United States, which are large net importers.
>
> **Price Expectations**
>
> When international oil prices spiked up to some US$75 per barrel in 2005 and 2006, there was widespread market speculation that the potential existed for prices to "superspike" to US$100 or more per barrel. These speculations were particularly strong in the financial industry and from oil traders. What in fact happened was that by the autumn of 2006 prices had fallen back to around US$60 per barrel.
>
> **Conclusion**
>
> There is no quantitative basis for confidently forecasting the development of international oil prices. The present historically high prices and the expectations for prices in excess of US$100 per barrel expressed earlier should not unduly color Chinese policy makers' consideration of energy security matters.
>
> *Source:* Authors.

virtual disappearance of spare production capacity. The principal reflection of this phase is that long-term crude oil futures prices were in the range of US$15 to US$25 per barrel in the 1990s and have now risen to the range of US$60 to US$75 per barrel. Some considerations affecting this new phase are presented in box G.2. For China, it would clearly be imprudent to plan energy policies that are based on the assumption that the strains on world oil supply will quickly disappear and that prices will fall back to the levels of the 1990s (that is, US$15 to US$25 per barrel). Instead, policy directions should be created that are sustainable at price levels as high as US$50 per barrel for extended periods.

Box G.2

Causes of the First 21st Century Oil Shock

Underinvestment

"The illusion that oil is in perennial oversupply has led to two decades of underinvestment in the oil industry. The world has been living off the legacy spare capacity built up many years ago." (Edward Morse, HETCO, as quoted in *Economist* 2005, 6)

Demand

In 2004, global oil demand grew at 3.9 percent, the fastest rate in more than 25 years (*Economist* 2005, 4). The increases fell back to 1.4 percent in 2005, still requiring the addition of 1.2 million barrels per day to world demand (IEA 2006). In December 2006, demand growth for the year was estimated at 1.1 percent or 0.9 million barrels per day, of which China accounted for nearly half while OECD demand fell by a marginal amount (IEA *Monthly Oil Market Report*).

Demand Outlook

The International Energy Agency in December 2006 estimated 2007 global oil demand growth at about 1.7 percent or 1.43 million barrels per day. China's consumption was expected to grow by 5.4 percent and contribute about 25 percent of the world increase, the same proportions as the Agency forecast for the Middle East. No other regions were expected to increase their consumption even half as fast as these two. Oil demand in Europe and in the OECD Pacific region was expected again to decline marginally. The forecasted global oil demand growth is equivalent to the production of three or four large oil fields (IEA *Monthly Oil Market Report*).

Supply

The demand shock reduced effective OPEC spare capacity in January 2006 to about 1.4 million barrels per day (IEA 2006). The IEA commented earlier that the last time that OPEC spare capacity was not a market issue was from 2000 to 2002, when that capacity was nearly 5.0 million barrels per day. At the end of 2006, as a result of relatively weak demand growth in that year, the IEA assessed effective OPEC spare capacity at 2.4 million barrels per day (IEA *Monthly Oil Market Report*).

Outlook for Spare Capacity

"While it is hoped that increasing investment will raise the level (from 1.3 million) [of spare capacity] over time, current non-OPEC investment schedules suggest this will not re-enter the comfort zone in the near future." (IEA 2005, 4).

Source: Authors.

Potential for Short-Term Interruptions of Oil Supply (Force Majeure)

For policy makers, the potential for short-term interruptions of oil supply tends to be the security issue of greatest concern because of the economic output loss and social disruption that can result. Fortunately, interruptions are not frequent. Interruptions or delays in oil flows could occur as a result of a variety of events taking place outside the importer's borders. Because of the integration of global oil and tanker markets, there can be adverse impacts on importers' interests—for example, as a result of price spikes or effects on trading partners' economies—even though there may be no direct impact on their oil import flows.

Interruptions could occur in many circumstances. One possibility is civil disorder in oil-producing areas, such as in Nigeria or the República Bolivariana de Venezuela. Another is a military action affecting major oil exporters, such as the Iran-Iraq War of 1981 to 1984 and the Iraq War of March 2003. Terrorist activities could affect tanker traffic in global oil chokepoints such as the Strait of Hormuz in the Persian/Arabian Gulf and the Strait of Malacca between the Indian and Pacific Oceans (significant incidents have not yet taken place). There is also the potential for natural disasters at home or abroad to damage critical oil and gas infrastructure. For example, in August 2005, Hurricane Katrina in the United States resulted in the temporary loss of about 1.2 million barrels per day of oil production (about 20 percent of the U.S. total) and a large amount of gas production. This loss occurred mainly because almost all offshore producing platforms in the Gulf of Mexico had to be shut down. Some of the platforms were damaged, and production of several hundred thousand barrels per day was still closed in mid 2006. Also, oil-refining operations were curtailed because units were shut down for safety reasons or were rendered inoperable by flooding or the loss of electricity supply. It should be pointed out that the hurricane also resulted in a reduction of U.S. oil consumption compared with expectations. Katrina was the worst, but not the first, hurricane to disrupt the U.S. oil industry. The experience of August 2005 highlights the importance of preparing for the unexpected in terms of security of oil and energy supply. Proponents of strategic oil storage in the United States and in the other member countries of the International Energy Agency probably never expected that the second emergency draw-down of strategic oil reserves (the first being during the Iraq-Kuwait crisis of 1990 to 1991) would be to deal with the losses caused by disruption of U.S. oil supply attributable to weather events.[3]

The world of the early 21st century has become a more dangerous place from the standpoint of securing uninterrupted supplies of oil in international trade than was the world of the last two decades of the 20th century. Some of the major oil-exporting countries appear to exhibit actual (Nigeria) or potential (Saudi Arabia) political instability. Significant tensions have arisen between the United States and certain exporters (the Islamic Republic of Iran and the República Bolivariana de Venezuela). Some large exporters (the Gulf states) may be militarily vulnerable vis-à-vis their larger neighbors. At the same time, the cushion of spare capacity to produce and transport oil that is available worldwide to cope with disruptions is proportionately smaller than in recent memory. Clearly, China must fashion its energy and oil policies with a view to reasonably minimizing oil import requirements, drawing its needed imports from a variety of sources, and taking steps nationally and internationally, by itself and cooperatively with other countries, to address the effects—direct and indirect—of any oil supply disruption that it may face.

Gas Supply Concerns

Gas supply is more reliable than oil supply, though concerns do exist.

Adequate Supplies at Competitive Prices

The global natural gas resource may be at least as large as the global conventional oil resource. However, it is developing more slowly, and the rate of gas reserve use is less than half the rate at which oil reserves are being produced. Much gas around the world is *stranded*, which means that there are resource discoveries capable of development but there are not facilities connecting the supply to markets. The gross supply situation in relation to actual and potential demands appears, on a global scale, to be better for gas than for oil, although the mature North American industry is facing problems, as discussed later.

It is difficult to generalize about gas prices in the same way as oil prices. Gas markets are regional rather than global. However, there seems to be agreement that it is technically possible to supply large volumes of gas in the form of LNG to major consuming areas at costs, including return on investment, in the range US$25 to US$30 per barrel of crude oil equivalent. The prices at which this gas is sold will, of course, depend on conditions in the various markets it serves.

A Good Record of Supply Security

Gas supplies, domestic and imported, in industrial countries have proven to be very secure, both for pipeline gas from local sources and for imported pipeline gas and LNG. Indeed, gas supply has, perhaps, been more secure than any other energy form. When security failures have happened, they have resulted mainly from technical accidents, supply shortages, and price disputes.

Technical Failures

In the case of interconnected transmission and distribution systems for gas supply, it is nearly always possible to mitigate the impacts of technical failures by cooperation within the industry. Such failures include pipeline rupture resulting from corrosion or mechanical damage or from natural disasters such as floods, earthquakes, or slope instability in mountainous areas. The most challenging technical failures are those in critical system components that take a long time to repair, such as large gas compressors and gas processing plants. An example of such a failure is the serious disruption caused by the Longford, Victoria, Australia, gas-processing plant incident in 1998 (box G.3). Gas supplied by a network of transmission

---

Box G.3

Regional and Liquefied Natural Gas Supply Issues

Australian Gas Supply Interruption, 1998

In 1998, a gas-processing plant explosion cut off all the gas supply to the state of Victoria (population 5 million). The resulting crisis was reported to have cost gas users more than US$1 billion.

Arun, Indonesia, 2002

The ExxonMobil LNG plant and export terminal, Indonesia's largest, was temporarily shut down because of security risks from separatist rebels in that area of Sumatra. Since early 2005, exports from Arun have been cut back below the level of contractual commitments because of continuing production problems in the area, despite the end of the insurgency there.

Skikda, Algeria, January 2004

An explosion destroyed three of six liquefaction trains and damaged the remaining three trains. In 2004, LNG production was reduced by 75 percent. Sonatrach, the owner of the plant, is rehabilitating the damaged trains and replacing the destroyed trains with one large new one, due on stream in 2007.

> **LNG Shipping**
>
> The Center for Liquefied Natural Gas (2005) claims, "The safety record of LNG ships far exceeds any other sector of the shipping industry."
>
> **The Russian Federation–Ukraine, January 2006**
>
> As the result of a dispute that appeared to relate to gas pricing, Russian exports to Ukraine were halted for four days and then quickly and completely restored. The root causes of the dispute are complex, and some are longstanding. Thus, prices were not market based, there were large payment arrears, Ukraine was alleged to have diverted gas intended for European Union (EU) markets, there was a dispute about ownership and refurbishment of transit pipelines, and there were issues about access to Russian gas stored in Ukraine for EU markets.
>
> **Supply Shortfalls**
>
> Concerns about the adequacy of gas supply from producing sources are rarely a short-term issue and then only when a single source dominates. Such a situation arose in the early 1970s in the Canadian province of British Columbia. The transmission pipeline system depended heavily on two producing fields. The fields were relatively new and had complex reservoir conditions. Water coning quickly developed as production rose, and a number of highly productive wells had to be abandoned.
>
> The immediate problem was solved by interrupting the supply to customers who did not have firm contracts and by securing alternative supplies to the affected markets. Subsequent exploration successes resulted in the development of new fields with much better producing characteristics, allowing the restoration of supply in two years.
>
> *Source:* Adapted from Stern 2006.

and distribution pipelines is, for engineering reasons, inherently a more stable and secure energy source than is electricity supplied under similar conditions.

Commercial Mechanisms to Secure Gas Supplies

Large gas buyers can help to secure long-term gas supply by due diligence. They can inquire about the suppliers' portfolios of gas and the alternative resources available to the suppliers. Carefully designed contracting with competent suppliers having a good track record is a well-tried way for buyers to secure a reliable supply.

Today in North America, there are concerns about the adequacy and price of long-term gas supply. These concerns persist even though, as this report is being completed, gas prices are low relative to crude oil, partly because of short-term demand effects mainly having to do with mild weather. After some 70 years of growth of the modern gas industry, it appears that the developed resource base is no longer capable of meeting demands and that the market will require supplemental supplies. Some technical, financial, environmental, and regulatory issues must be resolved in bringing on new large-volume sources, such as gas from Alaska and the Canadian north, coal-bed methane, and imported LNG. Some recent Canadian views, by government and by industry, as to North American gas supply security are summarized in box G.4.

In the meantime, prices are high by historical standards, and gas consumption appears to be falling slowly. For example, the industrial sector is using less gas. Manufacturers of ammonia and other fertilizers and some energy-intensive metallurgical industries have moved overseas to gas-rich jurisdictions such as Trinidad and Tobago. These developments are troublesome for policy makers and often result in ill-advised disruption of market functioning. If markets are allowed to work, investors will respond to higher prices and consider supply projects. If regulators promptly permit new investments, a better balance of supply and demand will be restored and prices can be expected to fall back, perhaps to LNG import parity.

Price Disputes

Price disputes in the gas subsector tend to be international rather than national and intergovernmental rather than intercorporate. Several such disputes arose in the 1970s and 1980s—between Canada and the United States, between Algeria and United States, and within Europe.[4] The disputes arose basically because pricing provisions in the contracts before the energy crisis of the 1970s did not allow gas prices to increase with the rapid rise in commodity value related to soaring oil prices. There were also differences in opinion about the methodology of determining netback prices. A gas dispute between the Russian Federation and Ukraine led to curtailment of supplies to Ukraine in January 2006 (see box G.3). The proximate cause was the fact that prices to Ukraine were very low compared with the prices that Russia obtained for exports to the European Union (EU) and no price-escalation mechanism had been agreed to.

Box G.4

Canadian Views on the Security of Gas Supply

By a Policy Maker

"Canada has an abundance of gas resources and the necessary infrastructure to deliver natural gas to market.

"Our market-oriented energy policy framework has been successful in providing us with a competitively priced, secure source of natural gas supply.

"Mature conventional North American natural gas basins, ever-increasing demand, and higher gas prices are attracting investment.

"New sources of natural gas supply—coal-bed methane, Alaska, the Canadian North, and methane hydrates—will arrive.

"Governments and regulators have a role to play in creating a favorable climate for investments in resources and infrastructure."

—Graham Flack, Associate Assistant Deputy Minister for Energy Policy, Natural Resources Canada

By a Natural Gas Industry Representative

"Current market tightness combines with environmental issues to produce perverse public attitudes respecting support for new supply.

"The public does not obviously share industry's perspective on importance of gas to our long-term energy future.

"Ensuring reliable policy support rests in good measure on reliable public support.

"Industry (and government) needs to articulate why gas passes the test of sustainability (in a multidecade context) and how we are working to ensure the industry's sustainability.

"[We] need a more systematic approach and need to get governments and other communities working with industry."

—Michael Cleland, President of the Canadian Gas Association

Source: Forum on Continental Energy, Woodrow Wilson Center, Washington, DC, March 21, 2005.

Very low gas prices cause resentment in the gas-producing jurisdictions, discourage the development of new supplies, and encourage buyers to increase consumption. The government of Canada used its energy regulator as the instrument to break the pricing provisions of existing export contracts and drive up prices almost in step with rising oil prices. Canadian disappointment with export pricing was one factor that led to a steady decrease in gas exports through the mid 1980s. The government of the Netherlands used diplomacy with its EU partners to achieve more reasonable prices for gas exports. Algeria and France engaged in intensive diplomatic dialogue with ups and downs but managed over time to resolve contentious issues. The dialogue between Sonatrach (Algeria) and El Paso and other importers (the United States) was not successful. The resolution of their differences ultimately took place in the U.S. court system.

A similar situation may exist today in relation to the export of Bolivian gas to Brazil and Argentina. There are clear parallels with the Russia-Ukraine situation. Bolivian export prices, which were determined by government-endorsed contracts, are much below commodity value in consuming areas in relation to today's oil prices. Bolivia is attempting, through discussions between regulators and also at a senior intergovernmental level, to achieve higher prices. It is possible to minimize price disputes and, if they arise, avoid harmful effects on supply by including appropriate pricing provisions in contracts, such as a provision for third-party dispute resolution. International experience strongly indicates that market-based deals negotiated between exporting and importing companies are much less prone to price disputes than are arrangements determined by governments.

Assessment and Implications

The rapid growth in the share of natural gas in energy balances worldwide is a positive factor, improving the security of international and national energy supply; providing a versatile, competitive energy source; and reducing environmental pollution. China, which is sharing in these trends, can make its gas supplies even more secure by diversifying sources to minimize technical risks, ensuring that contracted reserves are reliable, and insisting on freely negotiated contract terms that protect buyers' and sellers' interests in changing market circumstances.

Concerns about Electricity Supply Failures

Concerns about the electricity aspect of supply security arise primarily because of the North American and European blackouts in large, inter-

Table G.2 Some Major Electricity Failures of the Past 40 Years

Date	Region	Megawatts lost	Customers affected (million)
November 1965	New York and Ontario	20,000 (13 hours)	30.0
July 1977	New York	6,000 (26 hours)	9.0
1978	France	11,800 (several hours)	2.0
July 1996	Western United States	28,000 (up to 9 hrs)	
August 10, 1996	Western United States	Unknown	7.5
December 1997– January 1998	Ontario and Quebec, Canada, and parts of northeast United States (ice storm damage to transmission and distribution systems)	n.a.	1.4
August 14, 2003	Northeast United States and Ontario	60,000–65,000 (30 hours)	50.0

Source: Based on Hilt 2005.
Note: n.a. = not available.

connected but separately operated systems. They have affected large numbers of consumers in recent decades (see table G.2). There were extensive disruptions of power supplies in eastern Canada and the New England area of the United States. One set of failures resulted from the ice storm of the winter of 1997/98, which destroyed a high-voltage transmission line and took out many distribution lines when ice loads greatly exceeded their design parameters.

In 2003, another massive blackout occurred in the northeast United States and Ontario as a result of cascading failures caused initially by a tree falling on a power line. That incident and some other disruptions, such as a blackout in Italy during the summer of 2003, were blamed on the deregulation of the industry. That debate is beyond the scope of this report, but it should be noted that major blackouts occurred in New York and Ontario, Canada, in 1965 and in France in 1978 when utilities in those places were still vertically integrated. Those failures involved private utilities in New York and publicly owned ones in France and Ontario.

In most developing countries, supply disruptions occur mainly because of insufficient generation capacity attributable to flaws in the planning, approval, or execution of power projects, such as in China currently, or the lack of funding resources, as in Argentina, Brazil, and India during recent years. Supply disruptions of this kind have sometimes been the catalyst for profound reforms of the power subsectors in many large developing countries. The dimensions of major electricity failures occurring in industrial countries in the past 40 years are summarized in table G.2.

Concerns may also exist about fuel supply for electricity generation where the supplying industries have been affected by disruptions caused, for example, by labor disputes or by oil crises. There is no information showing significant impairment of electricity supply in any industrial country as a result of oil supply problems. In the United Kingdom, electricity supply was adequately maintained in 1984 and 1985, despite a coalminers' strike that lasted almost a year. The lack of impairment of electricity supply was attributable to the large coal stocks on hand when the strike began. Nuclear power was able to handle most of the base load, and oil-fired thermal stations absorbed a greater share of demand. This experience highlights the value of energy source diversification in the electricity industry.

A robust electricity supply system is an essential underpinning of a modern industrial state. As China's power system grows in size and complexity, and as reforms are introduced that enable the electricity subsector to capture the efficiencies associated with market behaviors and market pricing, policy will need to be directed to ensuring that investments are optimized in the various components of the industry and that operators maintain the highest standards. In this connection, valuable lessons can be learned from the experience of Europe and North America in avoiding power shortages and outages and in dealing with them when they occur.

Coal Supply, Competitiveness, and Environmental Concerns

Although international coal trade is long established and growing quickly, most of the world's coal tends to be used in the country that produces it. However, even the supply of coal drawn entirely from domestic sources may not be secure because of social disturbances or transport bottlenecks. It might also not be secure because of policies that focus on the economy's need for adequate and reliable sources of environmentally acceptable forms of energy at competitive prices.

During the past few years, railway bottlenecks have triggered important coal supply problems in China's coastal provinces. The supply problems have entailed extreme power shortages, especially during the summer. In the case of Western Europe, the old coal industries clearly failed to provide supplies that were adequate and competitively priced. Most of those industries have closed—including those in Belgium, France, Germany (hard coal), the Netherlands, and the United Kingdom—for this reason. There are also many examples of coal industries that did not meet the test of environmental acceptability and contributed disproportionately to problems of human health and loss of amenity associated with poor urban air quality in countries such as China and Turkey.

Dependence on a very large, domestic coal industry can cause energy security problems. Thus, adequacy of supply is not necessarily assured. Reliability may be impaired, for example, by railway transportation bottlenecks. Moreover, domestic supplies in all the markets served may not be internationally competitive, measured, for example, against the prices that Asian competitors pay for coal from Australia or LNG from the Middle East. China's policy makers therefore need to keep in mind the insecurity implications of excessive dependence on domestic coal.

Notes

1. The República Bolivariana de Venezuela may currently be contemplating such action (in this case, away from the United States and toward China).
2. The Mexican state company PEMEX is currently in that position.
3. The draw-down in connection with the Hurricane Katrina emergency was, however, small: about 11 million barrels out of a total reserve of some 685 million barrels.
4. European disputes arose between the Netherlands and the other European Union countries and between France and Algeria.

References

Center for Liquefied Natural Gas. 2005. "Updates on LNG Security and Safety." Center for Liquefied Natural Gas, Washington, DC.

Economist. 2005. "Oil in Troubled Waters." April 28.

Hilt, David. 2005. "Impacts and Actions Resulting from the 14 August 2003 Blackout." Paper presented at the North American Electric Reliability Council Research Teleseminar, January 20.

IEA (International Energy Agency). 2005. *Monthly Oil Market Report*. May.

———. 2006a. *Monthly Oil Market Report*. February.

———. 2006b. *Monthly Oil Market Report*. December. Accessible at http://www.oilmarketreport.org.

Stern, Jonathan. 2006. "The Russian-Ukrainian Gas Crisis of January 2006." Oxford Institute for Energy Studies, January 16. Accessible at http://www.oxfordenergy.org/pdfs/comment_0106.pdf.

APPENDIX H

Strategic Oil Reserves for China

This appendix summarizes and partially updates a paper prepared in January 2002 by the World Bank for the government of China. The paper represented a quick response to a request for an elaboration of the policy side of the strategic storage issue, an estimate of the possible costs of a strategic oil reserve for China, and an assessment of the economic impact of strategic storage as part of an economic stimulus package.

International Practice

Many oil-importing countries maintain strategic reserves as one element of civilian oil supply security policies. Germany, Japan, Switzerland, and the United States are examples of this approach. However, national policies are being superseded by the collective ones of the 26-member International Energy Agency (IEA) and the 25-member European Union (EU). The EU, which has maintained a stockpiling requirement since 1968, now requires its members[1] to maintain minimum stocks equivalent to 90 days of consumption of the three main groups of oil products, which account for 90 percent of oil use.[2] In contrast, the IEA's stocks requirement, which began in the 1970s, is to cover 90 days of net imports; hence, countries that are net exporters, like Canada and Norway, do not have a stockpiling obligation. Individual members of these organizations can, of course, hold larger stocks if they wish and, historically, Germany, Japan, and Switzerland have done so. Clearly, for countries with some domestic oil production, the EU standard requires larger stocks than the IEA standard. The U.S. Strategic Petroleum Reserve is the world's largest reserve. It has existed for more than 25 years, and in March 2006, it was equivalent to about 59 days of net oil imports by that country.

Commercial Stocks

Company stocks play a role in meeting these oil supply security goals, but competitive pressures encourage oil companies to minimize inventory costs. Commercial stocks, measured in terms of days of supply, have been trending downward for many years, although there are indications that the downward trend may have been exhausted by about 2000. Like the manufacturing industries, the oil industry has been moving toward just-in-time inventory practices. These practices mean that companies holding, for whatever reason, greater-than-average stocks will tend to be placed at a commercial disadvantage. There is no reason to believe that the companies factor in any provision for international supply disruptions in deciding on the level of their commercial stocks. Chinese oil companies, which are publicly committed to rationalize their operations and increase efficiency, will increasingly feel these competitive cost-minimization pressures. The task of providing security stocks cannot be left to the industry: governments have to play an active role, by mandating a certain industrywide level of commercial stocks (the EU approach), by holding strategic oil reserves themselves (the U.S. approach), or by doing both (the Japanese approach).

Rationale for Stockpiling

Strategic oil stocks, like other emergency oil supply measures, are maintained because of the importance of oil imports to the energy systems of many countries and the adverse effects on energy supply—and therefore on the economy at large—of interruptions in these imports. For example, the EU stockpiling directive states that any difficulty, even temporary, having the effect of reducing supplies or significantly increasing their prices in international markets could cause serious disturbances in the economic activity of the EU and that the EU must be in a position to offset or at least to diminish any harmful effects in such a case.

Rationale for a Particular Level of Stocks

The 90-day stock requirement can be related to expectations about the likely degree and duration of interruption of oil imports. The IEA requirement is intended to enable member countries to sustain economic activity unimpaired for three months even if all imports are lost, for six months if half of the group's imports are lost, and for one year if net imports are curtailed by 25 percent. The loss of 100 percent of net imports for 90 days implies a disruption of international oil trade on a scale that is scarcely conceivable. The effects of such a disruption on the oil-exporting countries would be so adverse that it is reasonable to assume that some measures would be taken within 90 days to at least partly restore supplies. A similar rationale about effects on exporting countries and likely corrective measures might be applied to scenarios in which net imports were reduced by 50 percent for six months or by 25 percent for one year.

Decisions to Draw Down Stocks

In the EU, the decision to draw down stocks would be made by the Council of Ministers. In the IEA, it would be made at the level of senior officials of member countries who comprise the Governing Board and who no doubt act on the instruction of their governments. Individual member countries of the EU and IEA are free to decide unilaterally to draw down stocks in emergency situations, always provided that this action does not result in failure to meet their international stockpiling obligations. Such decisions would be made at the most senior levels of government. In the United States, the president has the authority to draw down the country's Strategic Petroleum Reserve. President George H. W. Bush made that decision, in coordination with the IEA, for emergency reasons in 1990 and 1991, at the time of the Iraq War. In 2005, George W. Bush also decided to draw down the Strategic Petroleum Reserve to deal with shortages resulting from Hurricane Katrina. In relation to the size of the reserve, however, the draw downs were very small.

Costs

The cost of a strategic oil reserve comprises the cost of the oil—generally crude oil—assessed in 2000 at about US$20 per barrel (about US$70 or

US$80 per barrel in mid 2006, plus the cost of transportation), plus the cost of above-ground steel tanks in large concentrations, assessed in 2000 at US$25 to US$40 per barrel with a US$30 median estimate (likely significantly more in March 2006 because of the sharp rise in world steel prices). The cost of storage may be less, to the extent that converted oil tankers are used. Possible costs of underground storage were not assessed because they are site specific.

A Strategic Oil Reserve for China?

For 2000, assuming that an efficient Chinese oil industry would hold commercial stocks on the U.S. pattern, equivalent to about 50 days of consumption (this is a critical number that requires careful check), one would obtain the following results (the results for March 2005 are shown in parentheses):

- China could meet the IEA requirement of stocks to cover 90 days of net imports without a strategic reserve. (In 2005, China could still meet the IEA requirement with estimated commercial stocks equal to 106 days' net imports.)
- China would have to stockpile about 143 million barrels to meet the EU requirement of 90 days of consumption of major products, at a capital cost of US$6.5 billion to US$8.6 billion for oil and tankage. (In 2005, China would have to stockpile about 237 million barrels to meet the EU requirement, at a cost for oil alone of more than US$14 billion.)
- China would have to hold a strategic oil reserve of about 70 million barrels to match the U.S. Strategic Petroleum Reserve in relation to net oil imports, at a cost about half that of the preceding EU case. (In 2005, China would have to hold about 165 million barrels to match the U.S. Strategic Petroleum Reserve in relation to net imports.)
- By 2020, compared with 2005, the stockpiling requirements in the EU case would go up, pro rata to consumption, by about 65 percent. In the U.S. Strategic Petroleum Reserve case, pro rata to net imports, they would increase by a factor of 2.6. In the EU case, there would be a need to hold about an additional 120 million barrels of stocks.[3] Costs would, of course, increase by 2020, but it is not sensible to try attaching a figure to that increase because of uncertainties about construction costs and oil prices.

Quality of Estimates

The estimates in this paper are order of magnitude only. They and the underlying assumptions need to be tested in relation to Chinese conditions and experience.

Notes

1. New members (10 since 2000) are allowed a transition period to build up their stocks to meet this requirement. Members that are oil exporters (the United Kingdom) do not have a stockpiling obligation.
2. The three products groups are category I, gasoline-type fuels; category II, diesel-type fuels; and category III, medium and heavy industrial fuel oils (excluding ships' bunkers). At the end of 2005, the average coverage of stocks for each of the three categories was about 120 days.
3. These estimates are based on the following order-of-magnitude numbers for oil supply and demand in 2020 (2005 in parentheses) in millions of barrels per day: consumption, 11.0 (6.6); production, 3.8 (3.6); and net imports, 7.2 (2.8).

APPENDIX I

Predominant Approaches for Setting Regulated Tariffs for Gas and Electricity Transmission and Distribution

The method outlined in chapter 6 for setting regulated tariffs applies to transmission and distribution tariffs for both gas and electricity. Modern regulatory practice involves setting tariffs on the basis of a thorough analysis of costs in a tariff case every five years or so. In the intervening years, tariff adjustments are based on a prespecified formula tied to general cost changes (inflation) and achievement of performance standards, rather than on the company's own expenditures.

Regulated companies must adopt a proper system of accounts and accounting procedures that provide detailed and accurate financial, cost, and consumption data.[1] These data lie at the heart of the ability of an independent commission, such as the State Electricity Regulatory Commission, to regulate effectively the tariffs of all companies, taking into account factors that encourage economic use of the resources, good performance, and optimum investments. The data and reports that the company should provide as part of the full tariff filing include the following:

- The proposed aggregate revenue requirement, with detailed support for each element[2]
- The expected revenue at the prevailing tariff prices

- The proposed tariffs, terms, and conditions
- The expected revenue from the proposed tariffs
- Other information as required by the regulator.

The regulator should establish specific tariff filing requirements. The results of the regulator's investigations are enhanced if information is made available for public comment and scrutiny. A process of questions asked and answered should be initiated; transcripts of public hearings should be made available as should submissions whenever reasonably possible. Finally, the regulator should issue a written decision.

Period between Full Tariff Cases

The period between regulated tariff cases is usually set at three to five years. The longer regulatory lag gives the company opportunities to keep some benefits from efficiency-saving measures instead of passing them on immediately to customers. However, because prices will inevitably get out of line with costs over time, there will be pressure for review. Addressing the tension between the need to provide incentives for cost reduction and the need to maintain revenues in line with costs is one of the key issues in achieving good regulatory outcomes. Shorter periods reduce the risk that the tariffs get severely out of line with costs, whereas longer periods offer better incentives for cost reductions.

The standard length for most plans has been four to five years. However, a few early plans—notably the National Grid in the United Kingdom—have had shorter durations. The National Grid currently has a four-year plan. Alternatively, Victoria, Australia, used an eight-year duration for its initial plan. Scottish Power, Scottish Hydro, TransGrid, and EnergyAustralia all have five-year plans. The Australian National Electricity Code specifies a five-year minimum.

Three Simple Rules Necessary for Determining Tariffs

The recommended process for setting tariffs has three parts:

- Determine the required revenues of the company.
- Determine the form of the tariffs.
- Determine the rules for adjusting tariffs between tariff determinations.

Figure I.1 Three Simple Rules for Determination of Tariffs

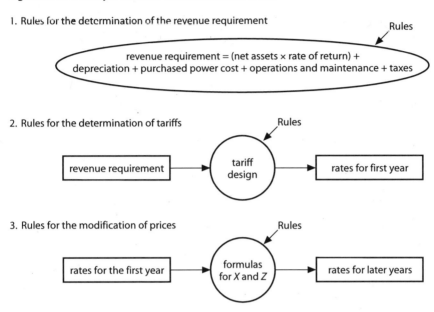

Source: [[AQ: Please provide.]]

This three-part framework restricts the company from raising prices above a certain threshold but allows it to keep the savings from efficiency improvements above a base level (represented by X in figure I.1), at least until the next scheduled tariff review. The framework also allows the regulated company to pass through certain costs (such as changes in taxation) over which it has no control. These are represented by Z in figure I.1.

Revenue Requirements

One of the key prerequisites for an efficient tariff reform involves an accurate definition of the company's revenue requirements—the total amount of revenue to which it is entitled. This amount is the basis for designing the tariffs.

This section reviews the issues involved in determining revenue requirements. On the whole, there is little dispute over the concepts of operating and maintenance costs and taxes. The main area in which there are varied methods is the determination of that part of the price reflecting the use of the capital employed in the business (that is, the return on assets).

There are three basic approaches to determining the appropriate level of costs:

- *Rate-base regulation* using the actual costs of the actual firm
- The *reference utility approach* involving replacement of service by a hypothetical efficient firm
- The *cost of an efficient firm* measured by the actual costs of comparable firms.

No method is perfect. However, a lot to can be said for choosing a method and sticking to it. Frequent changes introduce risk and increase the cost of capital to the regulated company.

Rate-Base Regulation

Traditional, U.S.-style rate-base regulation (also called *rate-of-return regulation*) sets the revenue requirement equal to the prudently incurred costs of the company being regulated. The assets that are included in the rate base are the physical assets that have been approved by the regulator, at the prices they actually cost, less depreciation. The revenue requirement consists of the following:

- Return on rate base (consisting of a mix of interest on debt and return on equity)
- Depreciation (or return of investment)
- Operating and maintenance expenses
- Taxes.

Rate-base regulation has been extensively used in North America; therefore, there is a lot of experience on its application and implementation. This method is important for private companies because they have a good chance of making an adequate rate of return on invested capital.

This type of regulation may require a lot of information exchange and control, which can be very problematic for the regulator, especially if it is a new regulator working in a situation where the required information has not been routinely collected and where an accounting system has yet to be set up. Nonetheless, the information required is necessary for any properly run business, and use of the rate-base method, combined with an incentive mechanism such as that described in the section titled "Tariff Adjustments between Major Tariff Proceedings" is recommended.

The Reference Utility Approach

Traditional in developing countries, the reference utility approach sets the revenue requirement equal to the theoretical revenues required by an optimally invested and managed utility serving the same territory as the one served by the real utility.

The major components of the revenue requirements are the following:

- Capital costs, set out on the basis of the reference grid
- Operational and commercial expenditures
- Taxes and other regulatory costs.

The *reference grid* is the optimum network (software based) designed to supply consumers according to the targets set out by the regulator (that is, technical quality). Typically, this network is valued at the replacement to new value, although in some cases historical asset costs have been used.

Operational and commercial expenditures normally are determined in two steps:

- Physical requirements (such as number of employees required to perform each process in the company; spare parts; and other requirements for operation, maintenance, and marketing) are determined, normally on the basis of benchmark-efficient companies.
- Physical requirements are priced at market values where the utility is located (that is, not international values but indigenous values).

This method has the advantage of making the regulator less dependent on information supplied by the utilities. It has the disadvantage of being difficult to implement.

This method is also criticized because it can produce returns that do not reflect a reasonable return on actual investments in the regulated utility. However, it produces strong incentives for the regulated utility to align investment and management decisions with those of an optimized company. The rationale behind this approach is that the regulator should not recognize revenues higher than those required by an optimized company operating in the same service area as the actual utility.

This method has been successfully implemented in many countries in Latin America and is beginning to be used in other places, such as Zhejiang province in China and several states in India. The choice was made in Latin America to make life easier for a new regulator without eliminating the efficiency and quality of service incentives. The method is very useful when accounting and financial data are not available or are inaccurate, as is the case in many emerging economies.

Comparable Firms (Top-Down Benchmarking)

Top-down benchmarking is usually implemented using costs of some benchmark company or group of peer companies. The problem is that no two companies' service territories, customer mixes, and growth patterns are the same, so determining what a reasonable level of cost is for a particular company by looking at the costs of other firms can be nearly impossible.

In cases in which there are many comparable firms operating in the same area, outcomes may be satisfactory (for instance, in Norway, where more than 200 electric cooperatives can be benchmarked).

Rules for Determining Tariffs

Having determined the total amount of revenue the regulated company is permitted to recover, one must calculate the tariffs that will determine from whom the revenues will be collected.

There are four steps in this process:[3]

1. Costs associated with the revenue requirement are apportioned to the different functions performed by the company (for example, transmission and storage).
2. Costs are classified as to whether they are fixed or variable. Fixed costs do not vary with the amount of the commodity transported (they include, for example, labor and capital-related expenses), whereas variable costs (such as fuel used to drive compressors) change with the facilities' use.
3. Costs are allocated between categories of network users. As the World Bank, Institute of Economic System and Management, and Public and Private Infrastructure Advisory Facility (2002, 284) state, "To achieve economic efficiency, it is generally accepted that fixed costs should be allocated to capacity as these costs are incurred, primarily to provide transmission services on the peak day. [In the case of natural

gas transportation] these will [usually] be the [urban gas distribution companies'] supplying . . . residential and commercial customers."
4. Tariffs are established for different groups of users. According to the World Bank, Institute of Economic System and Management, and Public and Private Infrastructure Advisory Facility (2002, 284), "A wide variety of possibilities of tariff design may be appropriate. . . . The most basic pricing concept would be to derive *postage stamp* prices such that customers pay the same price [tariff] regardless of location or distance from the source of gas supply. The most sophisticated approach, although subjective and theoretical, is receipt to delivery point pricing (distance based) on the basis of long-run marginal costs."

Tariff Adjustments between Major Tariff Proceedings

Between full tariff reviews (tariff cases), the company should be allowed to change tariffs by using an inflation index (the retail price index, or RPI) and an adjustment factor, known as X (the efficiency factor).

This framework has various names:

- Incentive regulation
- Performance-based regulation
- RPI _ X.

In theory, the factor X should be an efficiency index that is based on productivity increases. In practice, it is usually an estimate of expected changes in costs between rate cases. Cost changes for fuel and transport are often passed straight through to customers, so the system is sometimes known as (RPI _ $X + Z$), where Z is the allowance for permitted cost changes as they actually occur.

An adjustment may be made to tariffs to incorporate rewards for exceeding service quality standards or penalties for failure to achieve them.

Concluding Comments

It is normal practice during restructuring to make an independent valuation of the assets of the company being restructured. The choice of which value to use depends on a number of competing factors—efficiency, objectivity, consistency with legal precedent, ability to assure companies of the safety of their capital, ease of understanding, and consistency with underlying accounting and tax practices. But the biggest problem by far is changing existing tariffs in the process of restructuring. So it may be desirable, for one time only, to impute the value of the assets implied by the current tariff.

In other words, the current economic value of an asset is the present value of the stream of cash flows it will produce. Thus, the market value of the assets of a regulated company depends on the tariffs it will be allowed to charge over the life of those assets, net of expenses. In other words, the asset value depends on the allowed tariffs.

For example, China could keep electricity tariffs at or near their current levels by starting with the acceptable level of revenues and backing into the asset values implied by the resulting levels of revenue available for depreciation and return on investment. New investment would have to be added to the total asset value at actual cost for purposes of setting tariffs in order to provide sufficient revenues to make those investments possible. Eventually, the artificially valued assets would be fully depreciated, and the asset values on the company's books would reflect asset values determined in one of the three ways described earlier.

Notes

1. In addition, they must develop proper techniques to measure power losses, supply interruptions, voltage and frequency variations, and other parameters of power supply.
2. The elements of the revenue requirement are the operating and maintenance expenses of providing electric services, a fair return on investment in assets made by the company to provide those services (the *rate base*), depreciation of that investment, and taxes and other mandated costs.
3. This process is described in World Bank, Institute of Economic System and Management, and Public and Private Infrastructure Advisory Facility (2002).

Reference

World Bank, Institute of Economic System and Management, and Public and Private Infrastructure Advisory Facility. 2002. *China: Economic Regulation of Long Distance Gas Transmission and Urban Gas Distribution*. Washington, DC: World Bank.

APPENDIX J

Lessons from International Experience

Relevant Examples of Losses Derived from Unsound Energy Pricing

Most industrial countries began to rely on competition and market pricing for energy commodities following decades of experimentation with policies designed to regulate markets so as to achieve certain policy objectives (see box J.1 on p. 242). From the late 1950s, in Canada and the United States, markets for oil were managed to maintain crude prices at above world levels in order to protect indigenous producers. Then, beginning in the 1970s, the same markets were managed at below world levels in order to protect those countries' consumers from the high international prices that followed the oil crisis in 1973 and 1974. In many European countries, noncompetitive coal industries were protected from competition with oil and imported coal. In the power industry, tariffs have systematically been manipulated with very negative results: investment was low, and quality of service suffered.

The high costs and damaging results of these measures were recognized from the 1980s onward as a strong, new international consensus emerged in favor of energy policies that opened markets and relied on competition. This consensus is reflected in the energy policy positions recommended by international institutions such as the International Energy Agency (IEA) and the Asia Pacific Economic Cooperation. It is

Box J.1
Impacts of Selected Failed Policies

Post–World War II Policy

Protection of domestic energy producers, largely against lower-priced international (Middle East) oil and oil products, took the following forms:

- *The United States, 1957 to 1973—quotas on oil imports.* Oil prices exceeded world levels, international competitiveness was impaired, and regional and sectoral special interests were created.
- *Canada, 1961 to 1973—informal quotas on imports.* Uneconomic refineries were established and interregional tensions were created.
- *Europe, 1950s to 1990s—massive misguided investment in coal mines and power stations.* Social problems were deferred and made worse, industrial competitiveness was impaired, environmental damage became worse, transition to natural gas was slowed, and investments were made in the wrong "stock" of fuel-using appliances in households and industry.

Policy Following the 1973 to 1974 Oil Crisis

Protection of domestic oil consumers against the high prices of international crude oil and oil products took the following forms:

- *The United States, 1973 to 1980—domestic oil prices held below international levels.* Action increased demand, discouraged supply, worsened the domestic energy crisis of 1978 to 1980, and made adjustment to international levels from 1981 onward more difficult.
- *Canada, 1973 to 1985—same plan as the United States, except that deviations from the international price levels were much larger.* Such action caused interregional and intersectoral tensions (East versus West and producers' versus consumers' interests) and first deferred, then accentuated, inflationary pressures as Canadian prices caught up with international levels.

Source: Joskow 2006.

arguable that the general adoption of market pricing, particularly for oil and gas, enabled the IEA countries over the past 20 years to reap large economic efficiencies and to deal with the effects of the recent run-up in international oil prices much more effectively than they dealt with similar price increases in 1973 and 1974.

Policy makers have long recognized that the costs of resource misallocation resulting from noncompetitive energy pricing are high. There is less agreement when it comes to quantifying those costs. The following sections present some quantitative assessment of the costs of non-market-oriented energy pricing in the Organisation for Economic Co-operation and Development (OECD) countries.

Crude Oil and Products

It is generally agreed that noncompetitive oil pricing in the United States from 1973 to 1981 and Canada from 1973 to 1985 imposed heavy costs on the economy (see box J.1). The categories of increased costs included subsidies for noncompetitive energy forms (for example, U.S. coal gasification); penalties for competitive energy forms (conventional crude oil); distorted trade flows (for example, shipment of crude oil over uneconomic distances within Canada rather than exporting to the United States); protection of inefficient producers (for example, small oil refineries in the United States); prohibition of economic oil exports (for example, the prohibition restricting Alaskan oil from being sold outside the U.S. market); and the very high regulatory costs, including legal costs, of comprehensive oil price regulation in Canada and, particularly, the United States.

Natural Gas

A study carried out in 2001 has estimated that, compared with continuation of tight regulation, a policy of phased deregulation of the U.S. natural gas industry over the 23 years from 1977 to 2000 would have resulted in close to US$650 billion in savings of energy costs (Lemon 2001). Put another way, that number reflects the cost of nonmarket pricing under the assumption of continued tight regulation. Factored into the estimate of savings is a reduction in oil import costs associated with greater gas use in the deregulation scenario. Other benefits, such as improved environmental quality and reduced security concerns from oil import dependence, were not included in the estimated savings.

Another researcher examined the changes in consumer gas prices in the United States during the deregulation period.[1] He concluded that prices in the period from 1984 to 1994 had fallen by between 27 and 57 percent in real terms. Furthermore, box J.2 illustrates how, in a supply-constrained situation, competitive prices provided clear signals to reduce nonessential demand and triggered policy debates about long-term solutions to increase supply in the United States.

Electricity

The process of restructuring electricity markets has been more difficult than for crude oil and natural gas. This difficulty is, in part, attributable to the nature of electricity and the consequent complexity associated with its production and transmission.

Reviewing the U.S. experience recently, one eminent authority concluded that empirical evidence suggests that well-designed competitive market reforms have led to performance improvements in a number of dimensions and have benefited customers through lower retail prices. However, the transition to competitive electricity markets has been a difficult and contentious process. Although significant progress has been made on the wholesale competition, the framework for supporting retail competition has been less successful, especially for small customers (Joskow 2006, p. 33).

Joskow (2006) stresses that "many of the technical problems associated with creating well-functioning competitive electricity markets have been solved" but that "creating competitive wholesale markets that function well is a significant technical challenge and requires significant changes in industry structure and supporting institutional and regulatory governance arrangements." He continues (p. 33), "It requires a commitment by policymakers to do what is necessary to make it work. That commitment has been lacking in the U.S."

Effects on General Economic Performance of Nonmarket Energy Pricing

Most cases of nonmarket pricing relate to prices being held below internationally competitive levels for perceived public interest reasons. However, after the fact, such pricing policies almost always seem to have had perverse macroeconomic effects. For example, Canada held its oil prices below international levels from 1973 until 1985. Its European competitors had followed the international market throughout this period.

Box J.2

The Present Gas Supply-and-Demand Situation in North America

North American gas prices have been quite high since late 2000, but with a relatively warm 2006 winter and a warm start to the 2007 winter, prices have fallen well below their peaks. There have been sharp price spikes, most notably in the winter of 2000/01. Some industrial plants where gas is a very high component of total cost (ammonia fertilizers and methanol) have been shut down, or operations have been transferred to foreign gas-producing countries. These reports have naturally triggered reviews of the prevailing market functioning and pricing system by regulatory bodies. Those reviews had the following conclusions:

- With the important exception of some actions relative to the California market in 2000 and 2001, there have been no abuses of dominant positions in gas supply and pipeline transportation, and markets have worked satisfactorily.
- Market-determined prices played a major role in allocating gas supplies and using resources with maximum efficiency (short-term adjustments).

The events clearly show that the pricing system provided the right signals to consumers, producers, and policy makers to reduce nonessential demand and prevent disruption in the short term (demand side) and to take policy actions to attract investments in the longer term (supply side). After nearly 20 years of steady expansion, the industry needs new supplies to supplement the production from old fields that are now declining. These new supplies will come from deepwater offshore, coal-bed methane, liquefied natural gas, and frontier resources in Alaska and northern Canada. All of those sources are likely to be at a higher cost than existing supplies, and most involve large capital projects with long lead times, partly related to dealing with environmental regulations. Until the new supplies come on stream in large volumes, supply will be constrained relative to demands for additional uses, such as for clean electricity generation, and prices are likely to remain high (long-term adjustments in the absence of policy changes).

Producers and consumers have not moved to invite government intervention. Policy makers also appear satisfied that market pricing is still the correct solution for any problems of gas supply. However, it may be noted that the United States has enacted The U.S. Energy Policy Act of 2005 (see box 7.2). Title XVIII of the act allows for loan guarantees for the development of Alaskan natural gas resources.

Source: Authors.

Also, the United States reverted to international pricing in 1981 after eight years of pricing lower. Below-market pricing, Canadian politicians argued, would help keep inflation low and would give Canadians an energy cost advantage in international trade. Both arguments appear to be mistaken. With regard to inflation, Canada's performance measured over a long period was poorer than that of the United States, which had moved more quickly to international prices. With regard to international competitiveness, low energy prices in all probability led to Canadian producers being less energy efficient and therefore less able in the long term to deal with higher energy costs and remain internationally competitive.

Conclusions

It would require a lot of research and much new data to confidently quantify the cost of nonmarket energy policies and pricing in former centrally planned economies such as China. However, given experience in the OECD countries and on observation of conditions in China and countries of the former Soviet Union, we can confidently assert in qualitative terms that noncompetitive pricing and related resource misallocation has resulted in the following:

- Higher prices to the economy resulting from misallocation of resources at the production and consumption level, shortages, or oversupply (the latter two mainly in the power subsector)
- Reduced internal and international energy trade, resulting in loss of opportunities to reap comparative advantages from trade so as to optimize use of resources
- Less consumer satisfaction resulting from the absence of choice of suppliers and of competition among suppliers in terms of prices and services
- Low labor productivity in energy production, processing, and distribution resulting from a lack of competitive pressures to seek out operational efficiencies
- Unnecessarily high costs of regulation, both because competition in generation reduces the need for regulation and because light-handed regulation can be used to regulate transmission and distribution.

Note

1. Statement of Dr. Jerry Ellig, senior research fellow, Mercatus Center, George Mason University, before the U.S. House of Representatives Committee on Financial Services, Washington, DC, June 20, 2001.

References

Joskow, Paul L. 2006. "Markets for Power in the United States: An Interim Assessment." *Energy Journal* 27 (1): 1–36.

Lemon, Rodney. 2001. "California Here We Come: The Lessons Learned from Natural Gas Deregulation." Center for the Advancement of Energy Markets, Burke, VA.

APPENDIX K

Gas Price Formation and Gas Subsector Reform

Market Pricing Works

Prices matter a great deal in a socialist market economy. If they are allowed to work, they can progressively create huge efficiencies in economies transiting away from government-managed and government-controlled models. When prices are formed by competitive forces operating in a free market, they can be relied on to continuously balance supply and demand without government intervention. They help ensure efficient use of capital, enterprise, labor, and natural resources. They reflect consumer willingness to pay, thereby protecting consumers' interests more effectively than government regulation of price. Producers and consumers in many sectors of China's economy are benefiting from market prices.

Market Prices Can Work for Gas

Market pricing *can work* for the commodity natural gas in China: this has been demonstrated in many foreign jurisdictions. It generally *cannot work* for pipeline services that are used for the transmission and distribu-

---

This appendix is the executive summary on an unpublished gas price formation study carried out by the World Bank for clients in China during 2003 to 2004.

tion of the gas commodity, because they are natural monopolies. Modern economic regulation of transmission and distribution is therefore needed to protect consumer interests.

The Current Gas Pricing System Is Not Tenable

The present system in China causes inefficiencies, cannot work for gas imports, and is being eroded by the pricing flexibility allowed in new projects like the West-East Pipeline. Piecemeal approaches to fix it introduced further distortions, and a comprehensive overhaul of the system is required if the government's reform objectives are to be achieved. The challenges facing China have been met and overcome elsewhere. A great deal can be learned from international experience and consequences of ill-conceived government pricing intervention and the benefits of market solutions and sound regulation

The Target: Wholesale Competition

This model—implemented in North America and under way in Europe—focuses on achieving workable competition in the bulk supply of gas to large industrial consumers who are declared eligible to make their own gas purchase arrangements and to the urban gas distribution companies (UGDCs). It does not extend competitive choices to the retail level, but small consumers benefit from efficient gas purchasing by their UGDC suppliers. When the target is reached, (a) prices of the commodity will be determined by negotiated contracts and the competitive spot market, and (b) transmission and distribution tariffs will be determined according to a method approved by concerned authorities and will be regulated. Some services, such as shipping and storage, bundled with transmission and distribution at an early stage of the wholesale competition, could be unbundled and deregulated when conditions for acquiring these services competitively are met.

Transition Could Be Phased

A four-phase approach, with several intermediary steps is feasible. Figure K.1 (see p. 250) illustrates the four main components of a typical gas supply chain and indicates the contracting and pricing relationships in China prior to the recommended creation of competitive wholesale markets. Table K.1 (see pp. 251–53) sets out a phased approach to creation of a competitive wholesale market, which is described in more detail in the following paragraphs.

Figure K.1 The Gas Supply Chain: Structure, Contracting, and Pricing Prior to Introduction of Wholesale Competition

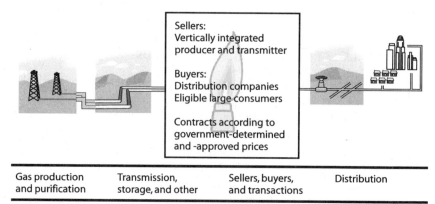

| Gas production and purification | Transmission, storage, and other | Sellers, buyers, and transactions | Distribution |

Price formation: Price of commodity and transmission is bundled and controlled by the central government, and the price of distribution is bundled with the generation and transmission price and controlled by provincial governments.

Source: World Bank study team.

Phase 1: First Steps—Establish the Preconditions

The preconditions for price negotiation between willing buyers and willing sellers of gas are identification of the exit of the purification plants (*gas-processing plants* in North American terminology) as the transaction point for a commodity of uniform quality, namely marketable pipeline gas (phase 1, step 1); introduction of price transparency in the various links of the supply chain (phase 1, step 2)]; and separation of gas supply from gas transmission (unbundling) and proper pricing and regulation of transmission services (phase 1, step 3).

Phase 2: Create a Framework for Negotiated Gas Prices

Gas commodity prices can be formed by negotiation, with regulatory supervision, even before strong supplier competition becomes effective. The starting point is a review of the method of current pricing, which will identify elements usable in the new approach (phase 2, step 1). One of the elements is market-value pricing, which can establish objective values at key points in the supply chain, on the basis of prices of competing fuels (phase 2, step 2.) Then the national oil companies' transmission operations should be organized as separate businesses for transaction and regulatory purposes (phase 2, step 3).

Table K.1 A Phased Approach to the Introduction of Wholesale Competition

Phases and steps	Description	Prerequisites
Phase 1	Meet preconditions for sound gas pricing and devise an adequate method for transmission and distribution prices and tariffs.	
Step 1	Map gas functions and identify transactions that require pricing: production, collection and purification (specification or pipeline-quality gas), transmission (including storage- and transmission-related services at least until market matures), and distribution. Confirm principles of modern or new gas pricing: commodity price (as defined at the outlet of purification plant) passthrough and transmission and distribution tariff, regulated on the basis of pricing policy approved by government. The exit (tailgate) of the gas-processing (purification) plant is clearly identified as the first point in the gas transaction chain.	None
Step 2	Define the following: • Upstream price at the purification plant outlet • Individual (pro forma in case of institutional constraints) accounts that are required for the separate functions to ensure transparency and adequate regulatory oversight of vertically integrated production and transmission functions prior to their separation. This step initiates price transparency.	None
Step 3	Separate commodity price at the purification plant outlet from gas transmission price. Distribution prices are already separate. Develop and approve method for transmission and distribution prices and tariffs (see World Bank and PetroChina 2004). This is the step of unbundling the tariffs for transmission and distribution services from the price of the commodity.	None

Outcome: Transparency is established for upstream price—transmission tariff, city-gate price, distribution tariff, and final selling price. Methods are agreed for transmission and distribution tariffs. Accounts of vertically integrated gas-supplying companies are separated to identify the production, transmission, and distribution functions. Through the disaggregation of components of the status quo, the informational preconditions are created for the move to negotiated commodity prices.

(continued)

Table K.1 (continued)

Phases and steps	Description	Prerequisites
Phase 2	Establish a framework for gas commodity price negotiation.	
Step 1	Review current method of commodity pricing, which might simply be government-set prices or might involve a degree of buyer-seller negotiation.	Successful completion of Phase 1
Step 2	Introduce market value techniques to set guidance prices: • Establish the value of gas as the weighted average of the prices of competing fuels by their market shares at the end use (basket-of-fuels approach). • Deduct regulated distribution tariffs and prices to determine city-gate price of the commodity. • Deduct regulated transmission tariffs and prices to determine pipeline-quality gas at the exit of purification plants. • Set the values of gas at outlet of purification and at city gates as benchmarks for producers and consumers.	Successful completion of Phase 1
Step 3	License separate transmission business units and apply regulated transmission tariffs for their operations (these tariffs have already been calculated in phase 2, step 2, for the purposes of establishing benchmark commodity prices.	Legal and corporate separation of transmission from production (even if ownership does not change)
Step 4	Establish parameter for gas commodity price negotiations and define regulatory oversight, thus paving the way for bilateral (buyer-seller) negotiation.	Successful completion of Phase 1
Step 5	Set limits on the ability of previous monopoly gas suppliers to retain supply exclusivities to large industrials and UGDCs and to monopolize access to pipeline capacity. This step clears the way for a progressively expanding "wedge" of negotiated pricing and supply arrangements to work into the total supply.	Successful completion of Phase 1

Outcome: There is agreement on the fuel-market value of natural gas at the end use, and by deduction of distribution and transmission tariffs from that market value, agreement on the value at the point at which pipeline-quality gas enters the system. These values provide benchmarks to adjust prices from the status quo level and to form a basis for later negotiation between sellers and buyers (phase 4). The transmission business units are separated from the operations of gas producers for operational and regulatory purposes, and the prices they charge are approved by the regulator. The scope for these "heritage" suppliers to influence or control the market is restricted by progressively reducing their rights to hold pipeline capacity and market share, thereby opening up access for other sellers, such as independent gas-marketing companies.

Table K.1 (continued)

Phases and steps	Description	Prerequisites
Phase 3	Extend commodity and transportation unbundling for UGDCs, initiate UGDC reform, and address issues of commodity pricing by UGDCs.	
Step 1	Overhaul the distribution system: • Develop and approve distribution (transport after city-franchise gate) tariffs and separate (unbundled) commodity price from distribution tariff. • Devise and implement measures to achieve social objectives if required.	Successful completion of Phase 2
Step 2	Regulate the gas price component of bundled distribution service, address-related policy, and social issues.	Successful completion of Phase 2

Outcome: Distribution tariffs are separated into distribution and commodity components, which increase transparency and prepare the way for direct sales of gas by producers or marketers to large customers who are served by UGDCs. The gas price to small consumers is regulated on a flow-through basis, and special measures such as lifeline rates are put in place to address social issues.

Phase 4	Complete the introduction of wholesale competition.	
Step 1	Increase the number of suppliers: allow importers and foreign partners in production-sharing contracts to sell their gas directly. This step clears the way for more suppliers to become active in the market.	Successful completion of Phase 3
Step 2	Establish open access to transmission systems.	Successful completion of Phase 3
Step 3	Allow participation of demand aggregators, marketers, traders, and brokers.	Successful completion of Phase 3
Step 4	Relax regulatory oversight of commodity negotiations as the market increasingly functions by itself and monitor for market power abuses. In the market for the gas commodity, government control has now given way to regulatory oversight to see to it that competitive forces are working freely.	Successful completion of Phase 3

Outcome: The market matures. Competitors enter the gas commodity business. Open access to gas transmission enables sales to be made directly to large consumers served off the transmission networks. Two-part UGDC tariffs enable sales directly to large consumers within the city gate. In the process of moving away from basket-of-fuels pricing, prices are set by negotiation between sellers and buyers. Regulatory oversight concentrates on monitoring rather than control of prices, but regulators intervene if dominant positions are abused. Small consumers are served by UGDCs purchasing bulk supplies at efficient prices in a competitive market and reselling to those consumers according to the commodity element of their tariffs.

Source: World Bank study team.

Most importantly, parameters would be established for probably bilateral price negotiations between suppliers and buyers to take place with a degree of regulatory oversight reflecting probably a weak state of competition (phase 2, step 4). Finally, supply exclusivities with time limits would be granted to encourage investment in new pipeline facilities (phase 2, step 5). Completion of phase 2 of gas price reform would eliminate the distortions that arise under present pricing, would provide valuable experience of price negotiation, and would enhance investor confidence, particularly with regard to badly needed new infrastructure.

Phase 3: Extend Unbundling

Unbundling should be extended to the UGDCs to separate gas supply costs from distribution costs and to identify (separately) efficient distribution service costs and gas purchase costs, both to be flowed through to captive customers under regulatory supervision, possibly with temporary government assistance to consumers for whom the costs of service of a modern, properly financed distribution operation may be initially burdensome.

Phase 4: Complete the Introduction of Wholesale Competition

The final steps to a functioning market with supplier competition involve the removal of any remaining constraints on gas sales by production-sharing contract partners (generally, the international oil companies) (phase 4, step 1); the formal establishment of nondiscriminatory open access in gas transmission (phase 4, step 2); the elimination of any remaining barriers to the participation of intermediaries such as traders and marketers (phase 4, step 3); and, finally, the withdrawal of regulatory oversight of the commodity market's functioning when competition can more effectively protect the interests of all market players (phase 4, step 4).

Table K.1 sets out in tabular form the four phases and the steps in each one. Figure K.2 shows what the gas supply chain would look like, in terms of pricing and contracting, after the introduction of wholesale competition.

Anticipated Outcomes

A competitive wholesale gas market will fully achieve all the theoretical and practical benefits of market pricing or the government's reform objectives. Economic efficiency will be optimized. In practical terms, shortages and surpluses will disappear, and the gas market will be self-

Figure K.2 The Gas Supply Chain: Structure, Contracting, and Pricing after the Introduction of Wholesale Competition

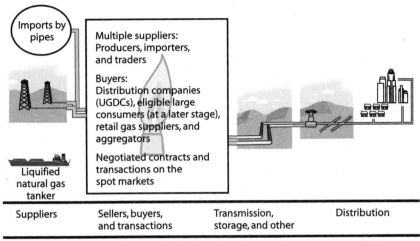

Price formation: Price of commodity is determined by negotiated contracts and the spot market and passed through the gas chain. Transmission and distribution tariffs are determined according to an agreed-upon methodology and are regulated.

Source: World Bank study team.

balancing. There will be no need for complex and expensive government interventions. Correct price signals will flow through to consumers and producers. These signals will be reflected in economically sound decisions on investment in production, purification, storage, transmission, distribution, ancillary services such as load balancing and gas exchanges, and gas-using equipment. Market pricing will enable gas to take its economically correct place in China's energy economy, if we assume, of course, that parallel steps continue to be taken to implement similar pricing principles in the markets for fuels that compete with natural gas. Government social or noneconomic objectives, if required, should be achieved through transparent and efficient direct (cross) subsidies with the least distortions of market (marginal cost-based) prices.

Reference

World Bank and PetroChina. 2004. *Gas Price Formation in China: Transmission Tariff Design.* Washington, DC: World Bank.

APPENDIX L

Pricing System to Support Adequate Implementation of State Council Document No. 5 on Power Subsector Reform

In the power industry, the 2002 State Council Document No. 5 was a major step forward. It clearly stated that the objectives of the reform in China are to continue the break up of the monopolistic structure of the industry and to gradually expand competition to improve its efficiency and ultimately provide the customers with the best service at the lowest possible cost. The plan details these objectives in eight points:

- Break up monopolies.
- Introduce competition.
- Increase efficiency.
- Perfect pricing mechanisms.
- Optimize resource allocation.
- Promote industry development.
- Promote the formation of a national grid.
- Establish a competitive electricity market.

The new policy sets out a number of major tasks to be achieved during the 10th Five-Year Plan (2001–05):

- Separate the transmission network from generation and restructuring of both generation and transmission network businesses.
- Establish competitive regional markets through the dispatch of generators according to bidding procedures and development of electricity market codes.
- Set up a new pricing mechanism that promotes environmental protection.
- Develop an efficient pricing mechanism for all parts of the electricity chain, including generation, transmission, distribution, and retail tariffs.

The plan envisages five types of pricing and implies that, ultimately, generation prices will be set by competition and passed through to consumers. This approach is entirely different from the one in place, which has just two tariffs—one for purchases from generators and one fixed-price tariff for sales to final customers. The following comments are from the 2002 plan itself:

- *Generator (network[1]) prices.* This term refers to prices for sale of electricity by generators into regional dispatch centers (the *grid price*). The grid price will consist of a government-determined fixed capacity price and a competitive market-determined energy price.
- *Transmission and distribution prices.* The government will determine the principles of transmission pricing and distribution pricing.
- *Retail tariffs.* The consumer price will be made up of the above prices and will change with the fluctuation of the grid price.
- *Transmission pricing for retail access.* Where there is direct supply to consumers, the (power) prices will be negotiated between generators and consumers, and the government will set the transmission price.
- *Emissions pricing standards.* These standards link the price paid to generators to their emissions so as to foster the development of renewable energy and other clean generation.

Most of the generation has been separated from transmission, but competition in the commodity market has not yet been established. Currently, all power is sold to the regional grid companies, which have a monopsony of purchases and a monopoly of sales. Five steps are required to make the markets competitive:

- Give large consumers (eligible consumers) and also distribution companies legal access to purchase freely from generators. Having a large number of customers is a prerequisite to efficient competition.

- Establish tariffs for the use of the transmission networks by the eligible customers and by the distribution companies. Such tariffs will give revenues to the transmission and distribution networks companies.
- Develop pool or spot-pricing (market) mechanisms to set market prices in real time. These spot prices will be used to clear the imbalances that naturally occur between contracted amounts and amounts actually generated for the contract or taken out of the grid by the consumer. Although this method is complicated, it is now well established.
- Establish a method for pricing contracts between distribution companies and generators (perhaps through auctions, cost of service, or bilateral contracting).
- Establish a method for retail tariffs that reflects fluctuating prices of the commodity, plus the transmission and distribution tariffs.

Bid-Price Pool

A major change will be the introduction of a bid-price pool. A bid-price pool will establish a (spot) market price for energy. Setting the price adequately will be critical.[2] Although the pool will set the price for spot sales[3] and purchases, most of the power will still be sold under contracts or tariffs. The market administrator will also use the spot price to determine the charges for the differences, or *imbalances*, between the contracts and the actual amount delivered. The spot price will also act as a reference price (as it is a price arising from the competitive market), thus affecting indirectly what buyers and sellers agree on for contract sales prices.

Environmental Costs

An additional item is the inclusion of prices for environmental factors. There are various ways to do so, and wide consultation and discussion among all concerned Chinese agencies are required to investigate these ways and decide how to incorporate environmental factors in the pricing system. Those prices have not been incorporated in table L.1.

Transmission Pricing

Transmission has been separated from generation—except for the generation assets that have been labeled as essential to the reliability and integrity of the system[4]—to prevent conflicts that can occur when the

Table L.1 New Tariff Requirements

Element	Who sells?	Who buys?	Who eventually pays?
Generation prices			
Contracts	Generating company	Large customers and distribution companies	Retail customers
Spot prices (in bid-price pool)	Generating company	Large customers and distribution companies	Retail customers
Transmission	Transmission companies	Large customers and distribution companies	Retail customers
Distribution	Rural and urban grids and departments at provincial network companies	Supply bureaus (and perhaps retail customers at a later stage)	Retail customers
Retail consumers	Retail tariffs for small customers should be made up of all the above items.		

Source: [[AQ: Please add.]]

entity in charge of transmission also generates its own power. There is an urgent need to develop separate transmission tariffs and to decide (a) what costs will be transferred to the transmission entity, (b) who will be paying for the entity's services, and (c) who will pay for investments. When eligible customers will be given a choice of supplier, they will make their own contracts with generators but pay other entities for the transportation. Predominant approaches used for setting regulated prices of network (transmission and distribution services) are described in appendix I.

One issue that perhaps needs explanation is how the (competitive) bid-price pool interacts with the contracts and with the (regulated) transmission prices. Detailed discussion of this issue is beyond the scope of this report.[5]

Final Consumer Pricing

It is necessary to distinguish between two types of consumers:

- Large consumers (who are allowed to choose their suppliers) will probably not be using the distribution system at all but will receive power directly off the transmission. They will pay separate transmission

charges to the transmission companies and will pay the generators for the commodity.
- For small consumers (who are not allowed to choose their suppliers), the distribution costs will be included in the final user tariff, which will include a passthrough of the transmission and commodity costs already paid by the distribution company. A clear, transparent, and workable mechanism for passthrough of generation and transmission costs into tariffs to final consumers must be designed and actually implemented.

Notes

1. The word *network* in the Chinese electricity sector refers both to the transmission network and to the tariff prices paid by the transmission operator to generators attached to the network. For clarity, we refer to "transmission network" and to "generator (network) tariffs."
2. Setting the price adequately will require sound market design and clear market rules.
3. Spot sales are for immediate delivery; contract sales are for later delivery.
4. This qualification is debatable, and some experts argue that all assets kept or acquired by transmission companies are not essential to the integrity of the system and that conflict-of-interest situations could arise in market operation.
5. These issues are discussed in detail in Hunt (2002).

Reference

Hunt, Sally. 2002. *Making Competition Work in Electricity.* New York: John Wiley & Sons.

Index

Boxes, figures, and tables are indicated by "b," "f," and "t" following page numbers.

A

accountability, 156
accounting procedures of regulated companies, 135
achievements in energy sector, 1–2
acid rain, 68, 70
adequate supply of reasonably-priced energy, 3–4b
advanced energy technologies
　decision-making about, 77–80
　11th Five-Year Plan and, 171, 174
　in Japan, 63–64, 79, 166
　necessity of, xl, 206–7
　potential of, xlv–xlvi, 83, 165–67
affordable housing, 60b
Air Pollution and Prevention Control Law of 2000, 75–76
air quality standards, 71
　See also environmental impact
Alaska National Wildlife Reserve, 215
Algeria, 91–92, 220b, 222

appliances
　direct burning, 188
　energy sustainability and, 150, 172, 173
　failed energy policies and, 242b
　less energy-intensive economic growth and, xxxviii, 43, 62, 62t
　lifestyle changes and, 15
　light bulbs, 49
Argentina, 93, 100, 225–26
Asia-Pacific Clean Development, 167b
Asia Pacific Economic Cooperation (APEC), 99, 112, 167b, 241
Association of Southeast Asian Nations (ASEAN), 112–13, 167b
Australia, 92, 112, 220, 220b, 227, 233
automobiles
　See also transportation
　electric vehicles, 59b
　growth in, 7, 54
　industrial policy and energy sustainability, 57, 58–59b
　manufacturing, 57
　ownership and sales, 55–56
Azerbaijan, 114

261

B

Beijing, 60b, 70
Belgium, 226
biodiesel, 188, 189t
biogas, 188, 189, 189t
biomass energy use, 187–91
  adverse effects on environment and
    human health, 188–89
  current status of development, 188,
    189t
  potential of, 188–89
  resources, 187
blackouts, 62, 88, 93, 115, 118, 224–25,
  225b
Blair, Tony, 80
Bonn Conference on Renewable Energy
  (2005), 97
BP, 100
Brazil, 25, 93, 114, 225–26
British National Oil Company, 100
building blocks for energy sustainability,
  xli–xlvii, liii, 157–69, 158f, 176–77t
  challenge of building block approach, 169
  cutting-edge technologies and transfer,
    development, and implementation
    (block 5), xlv–xlvi, 150, 165–66
  efficiency and sustainability (block 2),
    xlii–xliii, 150, 160–62
  institutions of energy policy (block 3),
    xlii–xliii, 150, 163–65
  integration within sound legal
    framework (block 1), xli–xlii,
    149–50, 158f, 159–60
  international cooperation (block 6),
    xlvi–xlvii, 150, 166–67
  mitigation of reforms on vulnerable
    groups (block 7), 150–51, 167–68
  mobilization of civil society and social
    organizations (block 8), xlvii, 151,
    168–69
  pricing and market fundamentals
    (block 4), xliv–xlv, 150, 165
buildings. *See* residential and commercial
  buildings

C

California Public Utilities Commission, 152
"California syndrome," 90, 96, 120b
Canada
  APEC and, 112
  blackouts in, 225, 225b

  effects on general economic
    performance of nonmarket energy
    pricing, 244, 246
  emergency use of spare capacity in, 106
  energy policy institutions in, 152b
  energy pricing in, 222, 242b, 243
  fuel allocation scheme in, 109
  gas supply in, 91–92, 221b, 222, 223b
  NOCs and, 100
  oil sands development incentive in, 98b
  power shortages in, 93
  strategic oil reserves in, 228–29
capital investment, 49
carbon capture and sequestration, 68, 77,
  80, 83
carbon dioxide emissions, 67, 68, 69, 72f
carbon finance for technology transfer,
  xlvi, 82–83b
CDM. *See* Clean Development Mechanism
cement industry, 52
central heating systems, 57, 60–61b, 61, 62
CERM (Coordinated Emergency
  Response Measures), 111b
certified emission reductions (CERs), 81,
  82–83b
Changshu 3F Zhonghao New Chemicals
  Material Co., 82b, 83b
Chevron-Texaco, 100
"China: Building Institutions for
  Sustainable Urban Transport"
  (World Bank), 58–59b
China Council for International
  Cooperation on Environment and
  Development, 78
China National Petroleum Corporation
  (CNPC), 102b, 103, 114
China Renewable Energy Scale-Up
  Program (CRESP), 97
city-saved scenario of urban
  transportation, 58–59b
civil society organizations (CSOs),
  168–69, 172
Clean Development Mechanism (CDM)
  advanced technologies and, xl, 166, 206
  policy emphasizing, 150
  potential of, xxxvii, 68–69, 79–81
  projects in implementation, 82–83b
  revenue stream available from, 207
climate change. *See* Kyoto Protocol
Climate Change Partnership, 167b
CNPC. *See* China National Petroleum
  Corporation

coal
  building energy use by, 1, 2, 5–6, 57
  clean-coal technologies, xlv, 67–68, 80, 166
    See also advanced energy technologies
  competitive market conditions and, 127–28, 141
  energy security and, 30, 94–95*b*, 226–27
  environmental impact of, xxxiii, 30, 67–69, 72–73, 226–27
  greener development, 75
  IGCC. *See* integrated coal gasification combined cycle
  leapfrogging technologies and, 72, 75
  power generation and, 28, 49, 50
  pricing, 133
    See also energy pricing
  production, 28, 30
  prospects though 2020, 28–29*b*
  reform agenda, 93, 94–95*b*
commercial buildings. *See* residential and commercial buildings
commodity markets, xiv, xlix, l, liii*t*, 108, 127, 133, 134, 171, 172, 177*t*
competitive markets, xiv, 2, 117, 127–28, 131–34, 138, 143–44, 241
consumption of energy. *See* energy consumption
cooking, pollution from solid fuel use in, 68
Coordinated Emergency Response Measures (CERM), 111*b*
CRESP (China Renewable Energy Scale-Up Program), 97
CSOs. *See* civil society organizations

D

deindustrialization, 31
DeLaquil, Pat, 78–79
demand-side management policies, 49–50
Denmark, 98*b*, 105
Development Research Center (DRC) of the State Council of China
  building energy efficiency, 37, 62
  coal industry and, 28–29*b*
  energy consumption projections, 11, 27, 44
  energy scenarios by, xxxiii, xxxvii, 7, 19–22, 40, 73–74

environmental strategy, 67, 73–74, 74*t*, 75, 76–77, 80
industrial sector and, 52–53
natural oil and gas resources, development of, 62
direct burning of biomass, 188, 189*t*
dissemination workshop feedback, 201–7
  attainment of target 20 percent energy-intensity reduction during 11th Five-Year Plan (feedback 3), 203–4
  challenges of present economy (feedback 1), 201–3
  revision of statistical system (feedback 2), 19, 202–3
  satisfaction of economic well-being, energy consumption, and environmental issues (feedback 4), 204
  use of leapfrogging technology (feedback 5), 205–6
domestic energy sources, xxxvii–xxviii, 92–93, 94–95*b*, 212–27

E

East-West Pipeline, 249
EEIA. *See* Energy Education and Information Agency
electricity
  biomass energy use and, 188, 189, 189*t*
  blackouts, 62, 88, 93, 115, 118, 224–25
  consumption and energy efficiency, 24–25, 49–52
  energy security and, 115–19, 121*b*
  generation capacity forecast, 7
  market restructuring, 244
  planning and maintenance procedures, improvement of, 115–18
  power generation, 47*t*, 49–52, 208–11, 209*t*, 210–11*f*
  power shortages, xxxii–xxxiv, 50, 51, 64, 93
    international concerns over, 224–26, 225*t*
  pricing, 133
    See also energy pricing
  renewable resources for generation of, 98*b*
  security of power systems, improvement of, 118–19
  separation of generation from transmission and distribution, 115, 116
  State Council Document No. 5, 143–44, 256–60

264 Index

electricity—*continued*
  sustainability, 174
  tariffs, 49, 133–36, 233–40, 235f, 259
electric vehicles, 59b
11th Five-Year Plan (2006-2010)
  comprehensive energy policy under, xxxv, 154–56
  energy efficiency, xlv, 64
  energy intensity under, 150
  energy-savings society under, 36
  environmental protection under, 75, 76
  new direction of, 41–43
  program under, xli, 162, 170–71
  20 percent reduction under. *See* 20 percent energy-intensity reduction
embargoes, 214
emergency allocation schemes, 109
Emergency Oil Sharing System (EOSS), 111b
Energy Charter Treaty, 114
Energy Commission
  beyond 11th Five-Year Plan, xlix–l, 172
  creation of, 149
  energy efficiency and, 150, 153, 172
  integrated and environmental policy and, xl
  role of, xliii, xlviii, 157–58, 159, 163
  sustainability and, 170, 172
energy consumption, 7, 11–34
  dissemination workshop feedback, 204
  11th Five-Year Plan, 36, 41–43, 203–4
  energy intensity, opportunity to reduce, 6–7, 35–66, 43–63, 44t, 45–47f, 47t
  energy statistics, 14–15, 19, 38, 192–97, 202–3
  gross domestic product (1980-2005) and, 13–14, 181–86t
  growth projections by, 19–22, 20t
  high growth forecast for 2000-2020, 20t, 21t
  key messages, 11–13, 35–37
  policy action, urgency for, 30–33, 31f
  primary energy consumption growth, 13–14, 14t, 16–17f, 37
  residential, 15
  sustained consumption growth, 1980-2000, 13–19, 18t
  10th Five-Year Plan, 22, 24–25, 37, 41
  trends in, 11–12
  unsustainable energy growth path, 14t, 21t, 22–24, 25t, 26t, 27f, 30–32, 31f

energy costs as proportion of GDP, 198–200, 199–200t
energy consumption and requirements compared, xxxv, 32, 44f, 44–45
energy intensity, 37–39
  increase in, 151
  stabilization of, 156
  10th Five-Year Plan and, 41
  trends in, xxxii, 15, 16f
Energy Education and Information Agency (EEIA), 150, 159, 161, 164
energy efficiency, 8, 35–66
  achievements in, 37–40, 38–40t
  avoidance of, 26
  dissemination workshop feedback, 203–4
  efficiency goal (building block 2), xlii–xliii, 160–62
  elements for 20 percent improvement program, 162b
  11th Five-Year Plan, 36, 41–43, 203–4
  energy intensity, opportunity for reduction of, 1, 43–63, 44t, 45–47f, 47t
  energy sustainability and, 5
  French Global Environmental Facility, 60b
  in hydrocarbon subsector, 2
  industry, 52–53, 53f, 54b
  key messages, 35–37
  policy and institutional framework improvement, 63–64
  power generation, 49–52
  products for, 49
  residential and commercial buildings and, 57–63, 60–61b, 62t
  target 20 percent energy-intensity reduction, 203–4
  10th Five-Year Plan, 41
  transportation and, 53–57, 55–56f, 58–59b
energy intensity
  achievements in reduction of, 1
  deindustrialization and, 31
  11th Five-Year Plan, 36, 42
  industrial countries, 25–26, 44, 46–47, 47t, 75
  measurement of, 161
  opportunity for reduction of, 43–63, 44f–47f, 47t
  policy to avoid highly energy intensive path, 12–13
  rise in, 12–13, 63
  10th Five-Year Plan, 31

Index 265

energy management services companies, 161
Energy Ministry. *See* Ministry of Energy
Energy Policy Act of 2005 (U.S.), 160b, 245b
energy policy institutions, xxxiv–xxxv, xliii–xliv, 95b, 150, 152b
energy pricing
  bid-price pool, 258
  in competitive markets, 2, 131–33, 138
  consumer pricing, 258–59
  development of sound strategy for, 129–31
  disputes in natural gas, 92
  energy commodity price policy, 141–45
  environmental costs, 258
  failed policies, lessons from, 241–47, 242b
  final consumer pricing, 259–60
  gas price formation and gas subsection reform, 248–55, 250f, 251–53t, 255f
  general economic performance and nonmarket energy pricing, 244, 246
  marginal cost of energy supply, 2
  monopolies, xlv, 128, 134, 142
  North American gas supply-and-demand situation and, 245b
  oil crisis of 1973 to 1974 policy, 242b
  post-World War II policy, 242b
  price controls, xlviii, 141–42, 170, 241–47
  reform of, xlv, 129–31, 143–44
  State Council Document No. 5, 143–45, 256–60
  tariffs, 133–36, 233–40, 235f
  transmission pricing, 258–59
Energy Research Institute (ERI)
  coal industry, 28–29b
  energy scenarios by, xxxiii, xxxvii, 7, 19–22, 40
  environmental impact, 67, 73–74, 74t, 75
  growth projections by, 11, 12, 19–22, 27, 44
  industrial sector, 52–53
energy security, 8, 87–125
  China's problems of, xxxii–xxxvi, 89–96, 90f
  choice of measures for, 119–21
  coal production, 94–95b
  definition, 2, 3–4b
  domestic energy sources, 92–93, 94–95b, 212–27
  electricity supply, 115–19, 121b
  emergency allocation schemes, 109
  foreign oil companies, security role, 101

  fuel-switching capabilities in large consuming industries, 108–9
  gas imports, 91–92
  geological factors and, 89
  international concerns and their causes, 90–93, 212–27
  international cooperation for short-term oil security, 110–15
  key messages, 87–89
  long-term security, strengthening, 120b
  measures for strengthening of, 120–21b
  multidimensional nature of, 93, 95–96
  oil and gas transport and refining, xxxiii, 91, 95–96, 96–114, 107–8
  political, military, and diplomatic factors and, 89
  rationing systems for oil products, 109
  repatriation of equity oil, 101, 102b
  short-term supply security, 103–7, 104t, 110–12, 120b
  sustainability of energy and, xxxviii
  underinvestment in, 214–17, 217b
energy statistics. *See* statistics, Chinese system for
energy supply
  electricity, 115–19, 121b
  supply failures, 224–26, 225t
  international security, 212–27
  natural gas, 91–92, 219–24
  oil
    emergencies, 87–88
    imports, 212–19, 213t
  short-term oil security, 110–12
energy sustainability. *See* sustainability
Energy Technology Policy (IEA), 3b
Eni S.p.A., 100
entrepreneurship, 2
environmental impact, 8, 67–86
  biomass energy use, 189–90
  carbon. *See* carbon capture and sequestration
  carbon dioxide emissions, 67, 68, 69, 72f
  carbon finance for technology transfer, 82–83b
  Clean Development Fund, 82–83b
  coal industry, xxxiii, 30, 67, 68, 69, 72–73, 226–27
  conservation efforts and taxation, 137–38
  dissemination workshop feedback, 204
  11th Five-Year Plan, 154–55
  of energy scenarios, 72–74, 74t
  externalities and, 138

environmental impact—*continued*
  greener development, 74–83, 84*f*
  growth scenario, 73*b*
  IGCC power generation. *See* integrated coal gasification combined cycle
  key message, 67–69
  mitigation of energy use and, 138–41, 140*t*
  nitrogen oxide and sulfur dioxide levels, 30, 67, 68, 73–74, 74*t*
  oil industry, 32
  pillars of energy sustainability and, xxxviii, 6
  pollution level concerns, xxxii, 69–72, 72*f*, 80
  State Council Document No. 5, 143–44, 258
  taxation and, xlv, 137–38, 139
EOSS. *See* Emergency Oil Sharing System
ERI. *See* Energy Research Institute
ethanol, 188, 189*t*
ethylene industry, 52
Europe
  blackouts in, 224–25
  carbon dioxide emissions in, 71, 72*f*
  coal industry in, 92
  deindustrialization and, 31
  energy consumption growth projection, 22, 44
  energy intensity in, 7, 31
  energy pricing in, 242*b*
  gas price disputes in, 222
  general economic performance and nonmarket energy pricing, 244, 246
  residential buildings in, 57
  Sudanese oil sales to, 102*b*
European Union (EU)
  diplomacy and trading strength use by, 99
  energy security and, 3–4*b*
  gas prices in, 222
  gas supply in, 92, 221*b*
  minimum stocks, maintenance of, 105
  strategic oil reserves of, 228–30
  zero-emissions coal technology in, 80, 166
exporting countries' policies, 91
See also specific countries
ExxonMobil, 99, 100, 108

F

Federal Energy Regulatory Commission (U.S.), 117
feedback from dissemination workshop. *See* dissemination workshop feedback
fertilizers, 222, 245*b*
final energy consumption growth, 14–15, 18*t*
Five-Year Plans. *See* 11th Five-Year Plan (2006-2010); 10th Five-Year Plan (2001-2005)
flags energy security scenario, 113*b*
force majeure events, 91, 218–19
former Soviet Union, 29*b*, 246
France
  blackouts in, 225, 225*b*
  coal industry in, 226
  energy efficiency in, 50
  fuel taxation in, 136, 137*t*
  gas supply in, 92
  Middle East oil and, 97
  NOCs and, 100
  oil consumption in, 28
freight transport, 15, 22, 35, 54–55, 55*f*
  See also transportation
French Global Environmental Facility, 60*b*
fuel economy standards, 54
fuel-switching capabilities for industry, 108–9
future concerns about energy sector, 2–5, 7, 11–34, 35–66, 70

G

gas. *See* natural gas
gasification of straw, 188
gasoline taxation, 137, 137*t*
GDP. *See* gross domestic product
geopolitical uncertainties, 88, 91, 219
Germany
  coal industry in, 29*b*, 226–27
  energy consumption in, 30
  gas supply and, 92
  oil embargoes and, 214
  renewable resources for generation of electricity in, 98*b*
  strategic oil reserves of, 104, 228, 229
Gleneagles Summit, 80
Global Environment Facility, 97
green growth. *See* environmental impact
gross domestic product (GDP)
  energy consumption in China (1980-2005) and, 13–14, 181–86
  energy costs as proportion of, 198–200, 199–200*t*

Group of Eight, 80, 92
Guangdong, 155
Guangzhou, 70

H

Heilongjiang province, 60b
Homuz straits, 106, 218
household appliances. *See* appliances
housing, affordable, 60b
Hurricane Katrina, 88, 90, 218
hydrofluorocarbons, 81, 82–83b
hydrogen-based fuel cells, 77

I

IEA. *See* International Energy Agency
IGCC. *See* integrated coal gasification combined cycle
Implementation Decree for the Statistics Law, 19
imported energy, 1, 2, 26–27
See also natural gas; oil
India
  advanced energy technologies in, 77, 81, 83
  bilateral international relations with, 114
  energy intensity comparison, 1
  power shortages in, 93, 225–26
  indigenous resource alternatives, 88
Indonesia, 112, 220b
indoor air pollution, 68
industrialization model of developing countries, 25–26
industrial sector
  energy-consuming businesses, targeting, 161–62
  energy efficiency and, 38–40, 39t, 52–53, 53f, 54b
  fuel-switching capabilities in, 108–9
  opportunity for reducing energy intensity, 43
insecurity of energy sector. *See* energy security
institutional capacity, 95b
integrated coal gasification combined cycle (IGCC), 51, 77–78, 83, 208–11, 209t, 210–11f
integrated policy within sound legal framework, 158f, 159–60
international cooperation, xl, 80, 110–15, 166–67, 171, 206–7
  See also Kyoto Protocol

International Energy Agency (IEA)
  carbon dioxide emissions, 71, 72f
  China and, 167b
  competitive markets policy, 241
  energy consumption and economic growth trends, 19, 25–26, 110–12
  energy security definition, 3b
  oil demand projections, 27
  oil importers and, 99
  short-term oil security, approach to, 111b
  strategic oil reserves and, 218, 228–29, 230
international oil companies (IOCs), 91, 97, 99–100, 101, 111, 112, 215, 218
Iran, 91, 219
Iran-Iraq War, 91, 218
Iraq-Kuwait War, 112
Italy, 92, 100

J

Japan
  advanced technology use in, 63–64, 79, 166
  APEC and, 112
  ASEAN and, 113
  diplomacy and trading strength use by, 99
  energy consumption in, 30
  energy costs as proportion of GDP in, 15, 32, 38, 44f, 45f, 45–47, 47t, 198–200, 199–200t
  energy demand in, 5, 28
  energy efficiency in, 50
  fuel taxation in, 57, 137
  gas supply and, 92
  mass transit culture in, 56, 57
  Middle East oil and, 97
  NOCs and, 100
  Northeast Asia Energy Cooperation and, 114
  strategic oil reserves and, 104, 228, 229
Japanese Ministry of International Trade, 79
Japan National Oil Company, 100
Jiangsu Meilan Chemical Group, 82b, 83b
Jinan Iron and Steel Group Corporation, 54b
Joint Oil Data Initiative, 111b
Joskow, Paul L., 244

K

Kazakhstan, 114, 152*b*
Kissinger, Henry, xxxix, 154
Korea, Democratic People's Republic of, 114
Korea, Republic of, 25, 29b, 92, 113, 114, 137
Kuwait, 91
Kyoto Protocol, xxxvii, 71, 82*b*, 166, 167*b*, 206, 207
 See also Clean Development Mechanism (CDM)

L

labor market, 48
Larson, Eric D., 78–79
leapfrogging technologies
 advanced technology systems and, xxxiix, 31, 37, 47, 48
 clean energy and, 8
 coal use and, 72, 75
 costs, 79
 definition, 205
 dissemination conference feedback, 205–6
 effect of, xxxvii, 54, 79
 energy efficiency and, 53
 energy sustainability and, 150, 163
 funding of, xvi, 4.2*f*, 83, 84*b*
 greener development and, 74, 75
 nuclear energy and, 75
 power generation and, 50
 technology choices and, xl
 transportation and, 53
legal framework and integrated policy, xl–xlii, 63–64, 149–50, 155, 158*f*, 159–60
Liaoning province, 60*b*
lifestyle issues, 7, 15, 48, 56, 204
lighting. *See* appliances
liquefied natural gas (LNG)
 reliability of, 91
 shipping, 221*b*
 terminal security, 6, 88, 90, 92, 215, 220*b*
low-income consumers, xlvi–xlvii, 168
low-trust globalization energy security scenario, 113*b*

M

Malacca straits, 106, 218
MARKAL model, 78–79
market-based reforms, 8–9, 127–47
 energy pricing in competitive markets, 131–33, 141–45
 key messages, 127–29
 mitigation of environmental effects of energy use, 138–41, 140*t*
 pricing and market fundamentals (building block 4), xliv–xlv, 150, 165
 reform agenda and pricing strategy, 129–31
 tariffs, 133–36, 233–40, 235*f*
 taxation of energy commodities, 136–38
mass transit. *See* public transportation
medium- and long-term energy conservation plan, 42
Mexico, 91, 112, 114
Middle East, 4*b*, 97, 99*t*, 106, 114, 227
 See also specific countries
military protection for crude oil traffic, 106–7
Ministry of Energy, 150, 163, 164*b*
Mongolia, 114
monopoly abuses, xiv, 128, 134, 142

N

National Development and Reform Commission (NDRC), 93, 94*b*, 152, 157
National Grid (U.K.), 233
national oil companies (NOCs), 91, 97, 99, 100–101, 102*b*, 111, 114, 214, 217
natural disasters, 88, 90, 91
natural gas
 assessment and implications for, 219
 commercial mechanisms to secure gas supplies, 221–22, 223*b*
 competitive market pricing system, 88, 143–44, 219–24
 deregulation of, 243–44
 development incentive for, 96–97, 98*b*
 diplomatic and trade relations and, 97, 99*t*
 gas regulatory commission, 171
 geopolitical uncertainties and, 88, 91
 imports, 91–92, 219–24
 NOCs and, 100, 101
 pricing, 92, 132–33
  controls, 128, 142–43
  disputes, 222, 224
  gas price formation and gas subsector reform, 248–55, 250*f*, 251–53*t*, 255*f*

supply issues, 219–20, 220b
tariffs, 133–36, 233–40, 235f
technical failures and, 220–21
transmission and distribution tariffs, 235f, 259
UGDCs. See urban gas distribution companies
wholesale market, 143–44, 248–55, 250f, 251–53t, 255f
NDRC. See National Development and Reform Commission
Netherlands, 92, 227
Nigeria, 91, 218, 219
9th Five-Year Plan (1996-2000), 13
nitrogen dioxide, 82b
nitrogen oxide and sulfur dioxide levels, 30, 67, 68, 69, 70, 73–74, 74t
North America
  See also United States
  blackouts in, 224–25, 225b
  energy consumption growth projection in, 22
  gas supply in, 93, 222, 223b, 245b
  lifestyle issues in, 7
Northeast Asia Energy Cooperation, 114
Norway, 228–29
nuclear power energy installation security, 6

O

OAPEC. See Organization of Arab Petroleum Exporting Companies (OAPEC)
OECD countries, 22, 44–45, 45f, 71, 72f, 246
  See also specific countries
offshore oil development, 97
offshore wind power, 83
oil
  commercialization markets, 88
  demand projections, 27
  development of national resources, 96–97, 98b
  diplomatic and trade relationships, 97, 98, 99t
  emergency supply, 87–88
  energy resources, optimum use of, 5–6
  energy security and, 96–114
  environmental impact, 32
  force majeure events, 91, 218–19
  foreign oil companies and security role, 101
  geopolitical uncertainties and, 88, 91
  imports, xxxii–xxxiv, 26–28, 32, 91, 95–96, 97, 98, 99t, 142, 212–19, 213t
  IOCs' role. See international oil companies
  maintenance of spare domestic oil capacity, 105–6, 217b
  mandated minimum commercial stocks, 105
  marginal conventional oil fields, 98b
  measures to strengthen energy security and, 120–21b
  military protection for crude oil traffic, 106–7
  monopolistic conditions and, 128, 142
  NOCs' role. See national oil companies
  normal commercial oil stocks contribution, 103
  offshore development, 97
  oil sands, 98b
  opportunity for reducing energy intensity, 43
  petroleum resource development, opening of, 2
  price controls, 128, 141–42, 241–47
  rationing systems for oil products, 109
  refineries, 43, 107–8
  repatriation of equity oil, 101, 102b
  security measures for oil supply, 88–89
  short-term security. See short-term oil supply security
  stocks owned by oil exporters in oil-importing countries, 105
  strategic oil storage, 103–4, 104t, 228–32
  taxation of refined petroleum products, 143
  2003–2006 price spike, 216b, 217b
  underinvestment in energy security for, 214–17, 217b
  unsustainable consumption trends, 30–31
oil-saved scenario of urban transportation, 58–59b
OPEC. See Organization of Petroleum Exporting Companies
open doors energy security scenario, 113b
Organisation for Economic Co-operation and Development (OECD)
  See also OECD countries
  on creation of IEA, 110
Organization of Arab Petroleum Exporting Companies (OAPEC), 91, 107, 214

Organization of Petroleum Exporting
  Companies (OPEC), 107, 213t, 214

P

Pacific region, 22
Pakistan, 114
passenger vehicle fleet, 55–56, 56f
  See also transportation
Pengfei Gao, 78–79
*The People's Republic of China Initial
  National Communication on
  Climate Change* (Government of
  China), 71–72
personal income, 15
Petro-Canada, 100
plant closures, xlvi, 168
policy banks, 161
pollution. *See* environmental impact
power generation. *See* electricity
pricing. *See* energy pricing
private investment, 95b
public transportation, 54, 55–56, 57

R

rationing systems for oil products, 109
regulatory mechanisms, xxxiv–xxxv,
  128–29, 139–41, 140t
Renewable Energy Promotion Law of
  2005 (REPL), 97
repatriation of equity oil, 101, 102b
reporting requirements for tariffs, 135–36
Repsol, 100
research and development (R&D),
  63–64, 80
residential and commercial buildings, 15,
  22, 48, 57, 60–61b, 61–63, 62t
road ahead scenario of urban
  transportation, 58–59b
Russian Federation, 1, 87, 91–92, 97, 99t,
  114, 221b, 222

S

Saudi Arabia, 91, 97, 105, 108, 110, 214, 219
Saudi Aramco, 99, 108
scenarios, xxxiii, xxxvii, 7, 19–22, 40,
  73–74
  city-saved scenario, 58–59b
  flags energy security scenario, 113b
  low-trust globalization energy security
    scenario, 113b
  oil-saved scenario, 58–59b

open doors energy security scenario, 113b
  road ahead scenario, 58–59b
Scotland, 233
security. *See* energy security
SEPA. *See* State Environmental Protection
  Administration
September 11, 2001 terrorist attacks, 90, 118
SERC. *See* State Electricity Regulatory
  Commission
service sector, 48
Shanghai, 70
Shell Group, 100, 112, 113b
short-term oil supply security, 103–7,
  104t, 110–12
  bilateral international relations and, 114
  global multilateralism and, 110–12
  maintenance of spare domestic
    capacity, 105–6
  mandated minimum commercial
    stocks, 105
  measures to strengthen, 120b
  military protection for crude oil traffic,
    106–7
  regional multilateralism and, 112–14
  stocks owned by oil exporters in oil-
    importing countries, 105
  strategic oil storage, 6, 103–5, 104t,
    218, 228–32
Siberia, 114
Sinopec, 108
social organizations. *See* civil society
  organizations (CSOs)
SOEs. *See* state-owned enterprises
solid fuel use in cooking, 68
Spain, 100
State Council Document No. 5, 143–44,
  256–60
State Electricity Regulatory Commission
  (SERC), xliii–xlliv, xlix, lii–liiit,
  128, 135, 163, 171, 176–77t, 233
State Environmental Protection
  Administration (SEPA), 71, 81
state-owned enterprises (SOEs), xxxvi,
  xlviii, 161, 170, 174
statistics, Chinese system for, 192–97
  current system and weaknesses, xlviii,
    38, 164–65, 193–94, 195–96t
  final energy consumption, 14–15
  groundwork in 1980s, 192
  recommendations for improvement, 19,
    194, 197, 202–3
  retrenchment in 1990s, 192

steel production, 52, 54b
strategic oil reserves, 6, 103–5, 104t, 218, 228–32
straw, gasification of, 188
strikes and blockades, 92, 214
Sudan, 102b, 114
Suez crises, 106
Sunda straits, 106
supercritical power plants, 51
sustainability
  See also building blocks for energy sustainability
  automobile industry and, 58–59b
  building blocks for sustainable path, liii, 157–69, 158f, 176–77t
  characteristics of comprehensive policy for, 154–56
  definition, 2, 4–6
  elements for 20 percent improvement program, 162b
  energy ministry, creation of, 164b
  opportunity for, 6–7, 12, 32–33
  pillars of, xxxvi–xxxiii, liii, 5–6, 89, 149–50, 156–57, 176–77t
  sequencing steps to, xlvii–l, liii, 170–74, 176–77t
  shift toward sustainable growth path, 7
  on threshold of change for, 151–54
Switzerland, 104, 228, 229

T

tanker transportation, military protection for, 106–7
targeted subsidies, 161
tariffs
  adjustments between major tariff proceedings, 239
  comparable firms, 238
  electricity, 49
  gas and electricity transmission and distribution, 233–40, 235f, 259
  period between full tariff cases, 234
  rate-base regulation, 236–37
  reference utility approach, 237–38
  regulation of, 133–36
  revenue requirements, 235–38
  rules for determining tariffs, 238–39
  State Council Document No. 5, 259t
  three rules for determining tariffs, 234–35, 235f
  top-down benchmarking, 238

taxation
  of energy commodities, 136–38, 137t
  fuel taxation, 129, 136, 137, 137t
  as incentive for energy reductions, xl, xlv, 171
  in Japan, 57
  of refined petroleum products, 143
10th Five-Year Plan (2001-2005)
  energy consumption trends, 14, 22–23, 37, 41, 42
  energy/GDP elasticity under, 151, 156
  energy growth, 24–25
  high energy intensity during, 31
  power generation, 50, 51
  reform agenda and, 40, 130, 131
  State Council Document No. 5, 256–57
  sustainability and, 151–52
  20 percent energy intensity reduction, 165
terrorist activities, 218
Texas Railroad Commission, 106
Tobago, 222
TotalFinaElf, 100
total suspended particulates (TSP), 69, 70
transportation
  electric vehicles, 59b
  emissions. *See* environmental impact
  energy consumption and, 15, 22, 35, 48, 54
  freight transport, 15, 22, 35, 54–55, 55f
  opportunity for reducing energy intensity, 43
  planning for energy efficiency, 7, 12, 53–57, 55–56f, 58–59b
  public urban strategy, 54–57, 58–59b
trifluoromethane, 81, 82–83b
Trinidad, 222
trucks, 55f. *See* transportation
Tsinghua University, 78
TSP. *See* total suspended particulates
Turkey, 25, 92, 227
Turkmenistan, 114
20 percent energy-intensity reduction
  action necessary to achieve, xliii, 161, 162b, 170–71
  data deficiencies and, 165, 202
  difficulty in achieving, 203–4
  importance of, 173
  step toward sustainability, xxxvi, 150, 154

U

UGDCs. *See* urban gas distribution companies

Ukraine, 87, 92, 221b, 222
ultrasupercritical power plants, 51
Umbrella Carbon Facility (World Bank), 82b
United Kingdom
    Canadian gas and, 97
    coal industry in, 29b, 92, 227
    marginal conventional oil fields development incentive in, 98b
    Middle East oil and, 97
    minimum stocks in, 105
    National Grid, 233
    NOCs and, 100
    oil embargoes and, 214
    tariffs, 233
United Nations, 167b
United Nations Framework Convention on Climate Change (UNFCCC), 71–72
United States
    alternative energy
        advanced technologies strategies and, 78–79, 166
        wind power development incentives, 98b
    APEC and, 112
    automobile culture in, 56
    deindustrialization and, 31
    Department of Energy. See U.S. Department of Energy
    electricity
        blackouts, 93, 224–25, 225b
        deregulation of, 116–17
        markets in, 244
        renewable resources for generation of, 98b
    energy costs as proportion of GDP in, 32, 44f, 44–47, 47t, 198–200, 199–200t
    energy efficiency in, 50
    energy intensity in, 31
    Energy Policy Act of 2005, 245b
    energy policy institutions in, 152b
    energy pricing in, 242b
        competitive markets, 117
        general economic performance and nonmarket energy pricing, 246
    EPA in, 81
    fuel-switching capabilities in, 108
    fuel taxation in, 136, 137, 137t
    gas
        natural gas deregulation, 243–44
        price disputes, 222
        supply, 92
    geopolitical tensions and, 219
    minimum stocks in, 105
    oil
        consumption, 28
        embargoes, 214
        imports, 28
        Middle East oil, 97
        noncompetitive oil pricing, 243
        pricing, 243
        strategic oil reserves, 104, 105, 228, 230
    residential buildings in, 57
    security of power systems in, 118
    Suez crises and, 106
    tariffs, 236
urban gas distribution companies (UGDCs), 142–43, 249, 252–53t, 254, 255f
urbanization, 48–49, 61
urban transportation strategy, 54–57, 58–59b
    See also transportation
U.S. Department of Energy (U.S. D.O.E.), 19, 79, 152b, 164b
U.S. Strategic Petroleum Reserve, 104, 105, 229, 230

V

Venezuela, República Bolivaiana de, 91, 106, 114, 218, 219
vertical coordination, 155–56
vulnerable groups, mitigation of energy reforms on, xlvi–xlvii, 167–68, 171

W

Wenying Chen, 78–79
wholesale gas market, 143–44, 248–55, 250f, 255f
wind power development, 81, 83, 98b
Working Group on Energy Strategies and Technologies (WGEST), 78–79
World Bank
    Clean Development Funds, role of, xlvi, 82–83b

coal development study, 93, 94–95*b*
CSOs and, 168
energy research of, 7, 57
power generation study, 49
renewable energy assistance from, 97
on space heating in buildings, 61*b*
on strategic oil reserves, 228
on tariffs, 238–39
Umbrella Carbon Facility of, 82*b*
urban transportation analyses by, 56–57, 58–59*b*
water and air pollution study, 68–69
wholesale gas market study, 144
World Trade Organization (WTO), 2, 143
Wu, Zongxin, 78–79

X

Xinjiang province, 97

Y

Yacimientos Petroliferos Fiscales, 100

ECO-AUDIT
Environmental Benefits Statement

The World Bank is committed to preserving endangered forests and natural resources. The Office of the Publisher has chosen to print **Sustainable Energy in China** on recycled paper with 30 percent postconsumer fiber in accordance with the recommended standards for paper usage set by the Green Press Initiative, a nonprofit program supporting publishers in using fiber that is not sourced from endangered forests. For more information, visit www.greenpressinitiative.org.

Saved:
- 11 trees
- 8 million BTUs of total energy
- 999 pounds of net greenhouse gases
- 4,148 gallons of waste water
- 533 pounds of solid waste